TC 7-100

Hybrid Threat

November 2010

DISTRIBUTION RESTRICTION:
Approved for public release; distribution is unlimited.

**HEADQUARTERS,
DEPARTMENT OF THE ARMY**

This publication is available at
Army Knowledge Online (www.us.army.mil);
General Dennis J. Reimer Training and Doctrine
Digital Library (http://www.train.army.mil).

Training Circular
No. 7-100

*TC 7-100
Headquarters
Department of the Army
Washington, DC, 26 November 2010

Hybrid Threat

Contents

		Page
	PREFACE	iii
	INTRODUCTION	v
	Hybrid Threats and the Hybrid Threat for Training	v
	The Emergence of Hybrid Threats	v
	Major Combat Operations	vi
	Multiple Threats	vi

PART ONE HYBRID THREATS

Chapter 1	HYBRID THREAT CONCEPTS	1-1
	Hybrid Threats	1-1
	Hybrid Adaptation	1-2
	Hybrid Transitions	1-3
Chapter 2	HYBRID THREAT COMPONENTS	2-1
	Threats and Other Actors	2-1
	Defining Enemy Combatants	2-3
	Use of WMD	2-7

PART TWO THE HYBRID THREAT FOR TRAINING

Chapter 3	HYBRID THREAT STRATEGY	3-1
	Strategic Operations	3-2
	Strategic Information Warfare	3-5
	Strategic Preclusion	3-6
Chapter 4	HYBRID THREAT OPERATIONS	4-1
	Operational Designs	4-1
	Regional Operations	4-2

Distribution Restriction: Approved for public release; distribution is unlimited.

*This publication supersedes FM 7-100, 1 May 2003.

i

Contents

	Transition Operations	4-2
	Adaptive Operations	4-3
	Principles of Operation versus an Extraregional Power	4-4
Chapter 5	**HYBRID THREAT TACTICS**	**5-1**
	Tactical Concepts	5-1
	Functional Tactics	5-3
Chapter 6	**HYBRID THREAT ORGANIZATIONS**	**6-1**
	Task-Organizing	6-1
	Military Organizations	6-1
	Insurgent Organizations	6-3
	Guerrilla Organizations	6-5
	Criminal Organizations	6-6
	Hybrid Relationships	6-8
Appendix A	**SCENARIO BLUEPRINTS**	**A-1**
	Scenario Blueprint Concept	A-1
	Exercise Design	A-1
	Scenario Blueprint Examples	A-2
	GLOSSARY	**Glossary-1**
	REFERENCES	**References-1**
	INDEX	**Index-1**

Figures

Figure 2-1. Combatant definitions 2-4
Figure 2-2. Paramilitary definitions 2-4
Figure 3-1. Strategic operations and other courses of action 3-2
Figure 4-1. Operational designs 4-1
Figure 5-1. Insurgent raid on a combat outpost (example) 5-4
Figure 5-2. HT area defense (example) 5-6
Figure 5-3. Insurgents provide security for a drug manufacturing site (example) ... 5-8
Figure 5-4. Guerrilla force conducts a raid (example) 5-10
Figure 6-1. Local insurgent organization (example) 6-4
Figure 6-2. Guerrilla hunter-killer company (example) 6-6
Figure 6-3. Large-scale criminal organization (example) 6-7
Figure A-1. MCO Blueprint COA sketch (example 1) A-3
Figure A-2. MCO Blueprint COA sketch (example 2) A-4
Figure A-3. MCO Blueprint COA sketch (example 3) A-5
Figure A-4. IW Blueprint COA sketch (example 1) A-6
Figure A-5. IW Blueprint COA sketch (example 2) A-7
Figure A-6. IW Blueprint COA sketch (example 3) A-8

Preface

The purpose of this training circular (TC) is to describe hybrid threats and summarize the manner in which such future threats may operationally organize to fight U.S. forces. It also outlines the strategy, operations, tactics, and organization of the Hybrid Threat that represents a composite of actual threat forces as an opposing force (OPFOR) for training exercises.

This publication applies to the Active Army, the Army National Guard (ARNG)/Army National Guard of the United States (ARNGUS), and the United States Army Reserve (USAR) unless otherwise stated.

Headquarters, U.S. Army Training and Doctrine Command (TRADOC) is the proponent for this TC. The preparing agency is the Contemporary Operational Environment and Threat Integration Directorate (CTID), TRADOC G-2 Intelligence Support Activity (TRISA)–Threats. Send comments and suggested improvements on DA Form 2028 (Recommended Changes to Publications and Blank Forms) directly to CTID at the following address: Director, CTID, TRADOC G-2 Intelligence Support Activity-Threats, ATTN: ATIN-T (Bldg 53), 700 Scott Avenue, Fort Leavenworth, KS 66027-1323. This publication is available at Army Knowledge Online (AKO) at http://www.us.army.mil and the Reimer Digital Library at www.adtdl.army.mil.

Readers should monitor those sites and also the TRADOC G2-TRISA Website at https://www.us.army.mil/suite/files/14705412 (AKO access required) for the status of this guide and information regarding updates. Periodic updates, subject to the normal approval process, will occur as a result of the normal production cycle. The date on the cover and title page of the electronic version will reflect the latest update.

This page intentionally left blank.

Introduction

This TC will address an emerging category of threats and activities that do not fit into the traditional understanding of conventional and unconventional war. It will focus on hybrid threats as simultaneous combinations of various types of activities by enemies and adversaries that will change and adapt over time. This TC summarizes the manner in which future threats operationally organize to fight us. However, it also discusses the strategy, operations, tactics, and organizations of the Hybrid Threat (HT), which portrays such forces in training exercises. For more detailed discussions of HT operations, tactics, and organizations, the reader should consult other TCs in the 7-100 series and supporting products.

HYBRID THREATS AND THE HYBRID THREAT FOR TRAINING

A *hybrid threat* is the diverse and dynamic combination of regular forces, irregular forces, and/or criminal elements all unified to achieve mutually benefitting effects. This introduction and the first two chapters will focus on the nature of hybrid threats that U.S. forces can expect to face in various operational environments. However, the remainder of this TC will focus on the representation of such hybrid threats in training exercises. In that context, the force that constitutes the enemy, adversary, or threat for an exercise is called the Hybrid Threat, with the acronym HT. Whenever the acronym is used, readers should understand that as referring to the Hybrid Threat, which is a realistic and representative composite of actual hybrid threats.

> *Note.* This introduction and chapters 1 and 2 address threats and use the terms *enemy* and *adversary* to refer to various nation-state or non-state actors that threaten or oppose U.S. interests. However, the remainder of this TC will use *enemy* and *adversary* to refer to an enemy or adversary of the actors who make up the Hybrid Threat (HT) for training exercises.

THE EMERGENCE OF HYBRID THREATS

The term "hybrid" has recently been used to capture the seemingly increased complexity of war, the multiplicity of actors involved, and the blurring between traditional categories of conflict. While the existence of innovative adversaries is not new, today's hybrid approaches demand that U.S. forces prepare for a range of conflicts. These may involve nation-state adversaries that employ protracted forms of warfare, possibly using proxy forces to coerce and intimidate, or non-state actors using operational concepts and high-end capabilities traditionally associated with states.

The emergence of hybrid threats heralds a dangerous development in the capabilities of what was labeled a "guerrilla" or "irregular" force in past conflicts. Hybrid threats can combine state-based, conventional military forces—sophisticated weapons, command and control, and combined arms tactics—with attributes usually associated with insurgent and criminal organizations. Hybrid threats are characterized by the combination of regular and irregular forces. Regular forces are governed by international law, military tradition, and custom. Irregular forces are unregulated and as a result act with no restrictions on violence or targets for violence. The ability to combine and transition between regular and irregular forces and operations to capitalize on perceived vulnerabilities makes hybrid threats particularly effective. To be a hybrid, these forces cooperate in the context of pursuing their own internal objectives. For example, criminal elements may steal parts for a profit while at the same time compromising the readiness of a U.S. force's combat systems. Militia forces may defend their town or village with exceptional vigor as part of a complex defensive network. Some hybrid threats will be a result of a state (or states) sponsoring a non-state actor.

Hybrid threats will seek to use the media, technology, and a position within a state's political, military, and social infrastructures to their advantage. In combat with U.S. forces, their operations can be highly adaptive, combining conventional, unconventional, irregular, and criminal tactics in different combinations that shift over time. They will use insurgent activities to create instability and to alienate legitimate forces from the population. Additionally, they will use global networks to broadcast their influence. Hybrid threats often will not place limits on the use of violence.

The phenomena of irregular forces engaging regular military forces by using conventional tactics and weapons is not new; the American Revolutionary War and the Vietnam War contain examples of pitched battles between regular and irregular forces. To an increasing degree, hostile groups are using advanced weapons, off-the-shelf technology, combined arms tactics, and intensive training to prepare their forces to engage U.S. Army troops when conditions are suitable. They are also perfectly capable of using terrorism and guerrilla tactics when that suits them. Whether these forces fight as an indigenous resistance or as a proxy for a hostile nation-state, some combination of these approaches appears likely in the persistent conflicts of the near future.

MAJOR COMBAT OPERATIONS

Major combat operations (MCO) employ all available combat power (directly and indirectly) to destroy an opponent's military capability, thereby decisively altering the military conditions within the operational environment. MCO usually involve intensive combat between the uniformed armed forces of nation-states. Hybrid threats may have the capacity to engage in MCO. Even then, MCO tend to blur with other operational themes. Within a theater of war, some U.S. or coalition forces may be conducting MCO while others may be conducting counterinsurgency and limited intervention. For example, in Vietnam both the United States and North Vietnam deployed their national armed forces. Although major battles occurred, the United States characterized much of the war as counterinsurgency operations.

MULTIPLE THREATS

Multiple threats to U.S. interests exist, and rarely are only two sides involved in modern conflicts. The potential for armed conflict between nation-states remains a serious challenge. Additionally, the influence of non-state actors has ever-increasing regional and worldwide implications. Some regional powers aspire to dominate their neighbors and have the conventional force capabilities to do so. Such situations may threaten U.S. vital interests, U.S. allies, or regional stability. Transnational groups conduct a range of activities that threaten U.S. interests and citizens at home and abroad. Such activities include terrorism, illegal drug trading, illicit arms and strategic material trafficking, international organized crime, piracy, and deliberate environmental damage. Extremism, ethnic disputes, and religious rivalries can also further the threat to a region's stability. Collectively, these transnational threats may adversely affect U.S. interests and possible result in military involvement.

PART ONE

Hybrid Threats

Part one of this TC focuses on simultaneous combinations of various types of activities by enemies and adversaries that will change and adapt over time. It summarizes the nature of such threats, their component forces, and the manner in which they would operationally organize to fight us.

Chapter 1

Hybrid Threat Concepts

A *hybrid threat* **is the diverse and dynamic combination of regular forces, irregular forces, and/or criminal elements all unified to achieve mutually benefitting effects.** Understanding hybrid threats involves several key concepts, most of which are not actually new.

HYBRID THREATS

1-1. Hybrid threats are innovative, adaptive, globally connected, networked, and embedded in the clutter of local populations. They can possess a wide range of old, adapted and advanced technologies—including the possibility of weapons of mass destruction (WMD). They can operate conventionally and unconventionally, employing adaptive and asymmetric combinations of traditional, irregular, and criminal tactics and using traditional military capabilities in old and new ways.

1-2. Threats can challenge U.S. access—directly and indirectly. They can attack U.S. national and political will with very sophisticated information campaigns as well as seek to conduct physical attacks on the U.S. homeland.

1-3. It is important to note that **hybrid threats are not new**. History is full of examples of how an adversary has prepared to use his relative perceived strengths against his opponent's perceived weaknesses:

- 1754 to 1763: regular British and French forces fought each other amidst irregular Colonialists fighting for the British and American Indians fighting for both sides.
- 1814: Peninsula War ended after the combination of regular and irregular allied forces from Britain, Portugal, and Spain prevented France from controlling the Iberian Peninsula.
- 1954 to 1976: Viet Cong and People's Army of Vietnam combined irregular and regular forces in fighting the French and U.S. forces. Viet Cong would organize into conventional and unconventional units.
- 2006: Hezbollah mixed conventional capabilities (such as anti-armor weapons, rockets, and command and control networks) with irregular tactics (including information warfare, non-uniformed combatants, and civilian shielding). The result was a tactical stalemate and strategic setback for Israel.

1-4. The U.S. Army will face hybrid threats that simultaneously employ some combination of regular forces, irregular forces, and/or criminal elements, to achieve their objectives. **Hybrid threats will use an ever-changing variety of conventional and unconventional organizations, equipment, and tactics to create multiple dilemmas.**

1-5. Hybrid threats seek to saturate the entire operational environment (OE) with effects that support their course of action and force their opponents to react along multiple lines of operation. A simple military attack may not present enough complexity to stretch resources, degrade intellectual capacity, and restrict freedom of maneuver. Instead, hybrid threats can simultaneously create economic instability, foster lack of trust in existing governance, attack information networks, provide a captivating message consistent with their goals, cause man-made humanitarian crises, and physically endanger opponents. Synchronized and synergistic hybrid threat actions can take place in the information, social, political, infrastructure, economic and military domains.

1-6. Opponents of hybrid threats will have difficulty isolating specific challenges. They will be forced to conduct economy of force measures on one or more of several lines of operation. Meanwhile, hybrid threats will continue to shift effort and emphasis to make all choices seem poor ones.

1-7. Hybrid threats are networks of people, capabilities, and devices that merge, split, and coalesce in action across all of the operational variables of the OE. Each separate actor and action of a hybrid threat can be defeated if isolated and the proper countermeasure is applied. By creating severe impacts across the total OE, a hybrid threat prevents its opponents from segregating the conflict into easily assailable parts. Often military action will be the least important of a hybrid threat's activities, only coming after exploitation of all the other aspects of the OE has paralyzed its opponent.

1-8. Hybrid threats can include criminals and criminal groups used in conjunction with both regular and irregular forces. A picture of this future was provided by the 2008 Russian-Georgian conflict, in which Russia employed the many criminal elements operating in South Ossetia to conduct the cleansing of ethnic Georgians from that region. Additionally, criminal organizations have the potential to provide much-needed funding to operations and facilitate the purchase of equipment. Adversaries will be enabled by WMD and technologies that allow them to be disruptive on a regional and area basis.

1-9. Swift tactical success is not essential to victory. The dimension of time favors those fighting the United States. An enemy need not win any engagement or battles; the enemy simply must not lose the war. Wearing down the popular support for U.S. operations by simply causing a political and military stalemate can be all that is required to claim victory or to change U.S. behavior or policy.

1-10. The most challenging attribute of our adversaries will be their ability to adapt and transition. Their speed, agility, versatility, and changeability are the keys to success in a fight against a larger, more powerful opponent.

HYBRID ADAPTATION

1-11. Adaptation, broadly defined, is the ability to learn and adjust behaviors based on learning. Adaptation is closely linked to one's OE and its variables. **Adversaries can approach adaptation from two perspectives: natural and directed.**

1-12. *Natural adaptation* occurs as an actor (nation-state or non-state) acquires or refines its ability to apply its political, economic, military or informational power. Natural adaptation may be advanced through—
- Acquisition of technology, key capabilities, or resources (financial and material).
- Effective organization.
- Effective use of the information environment or even key regional or global alliances.

1-13. *Directed adaptation* refers to adaptation, based specifically on lessons learned, to counter U.S. power and influence. Counters to U.S. actions will be ever changing and likely conducted by a hybrid force. Hybrid threats will offer a mix of capabilities along the spectrum of conflict to counter U.S. military actions. Adversaries will learn from U.S. operations what works and what needs refinement. They will be whatever the U.S. force is not. Like natural adaptation, directed adaptation will inform issues of force design, military strategy, and operational designs.

1-14. Success goes to those who master the skills necessary to act, react, and adapt with speed and creativity. Enemies learn quickly and change, often unconstrained by rules or bureaucracy. While this may cause haphazard and incomplete change, it does allow a rapidity that is difficult to counter. Adversaries will continue to be adaptive in terms of using all available sources of power at their disposal.

HYBRID TRANSITIONS

1-15. One of the most dangerous aspects of a hybrid threat is the ability of its components to transition in and out of various forms. Military forces, for example, can remove uniforms and insignia and other indicators of status and blend in with the local population. Insurgent forces might abandon weapons and protest innocence of wrongdoing. Criminals might don the accoutrements of a local police force in order to gain access to a key facility.

1-16. Hybrid threats will use the difficulties of positive identification of threat actors *as* threat actors to their advantage. OEs will be replete with many actors conducting activities counter to U.S. interests but without a clear visual signature as to their status as threats. Indeed, often these actors will be providing signatures similar to friendly or neutral actors.

1-17. Time-honored concepts of "conventional" and "unconventional" war and "traditional" methods versus "adaptive" methods are weapons to a hybrid threat. These concepts do not have meaning to a hybrid threat beyond their ability to be used against its opponents. Hybrid threats see war holistically and do not try to break it up into convenient pieces.

1-18. Hybrid threat forces will need to perform certain functions in order for them to succeed. Some functions at some points will best be performed by uniformed military forces. At other times or for other reasons, some functions will be best performed by irregular forces. At some points, both types of forces will be acting together. At others, they will shift between the status of regular and irregular. They may also use deception to shift between combatant and noncombatant status. Hybrid threats will present themselves in many ways but always maintain the ability to aggregate at the time and place of their choosing.

This page intentionally left blank.

Chapter 2

Hybrid Threat Components

Through formal structure and informal agreement, military and state paramilitary forces can work in concert to varying degrees with insurgent, guerrilla, and criminal groups towards common ends. Typically, the common goal is the removal of U.S. and coalition forces from their area of operations. The goals of hybrid threat forces may or may not coincide with those of other actors in the same geographic area.

THREATS AND OTHER ACTORS

2-1. There are many types of actors or participants in today's complex world environment. Some of the actors are countries (also called nation-states) and some are not. Nation-states are still dominant actors. However, some power is shifting to nontraditional actors and transnational concerns. There are many potential challenges to traditional concepts like balance of power, sovereignty, national interest, and roles of nation-state and non-state actors.

2-2. Of course, not all actors are threats. To be a threat, a nation or organization must have both the capabilities and the intention to challenge the United States. The capabilities in question are not necessarily purely military, but encompass all the elements of power available to the nation or organization.

2-3. Defining the actors in hybrid threat operations requires a dynamic situational awareness of change in a particular operational environment (OE). An order of battle or an appreciation of adversaries may transition abruptly or retain characteristics over an extended period. Similarly, the full band of PMESII-PT variables requires constant estimation and analysis to project or confirm the motivations, intentions, capabilities, and limitations of a hybrid threat. This section addresses significant categories of threats that can combine, associate, or affiliate in order to threaten or apply hybrid capabilities.

2-4. The key components of a hybrid threat, therefore, are two or more of the following:
- Military force.
- Nation-state paramilitary force (such as internal security forces, police, or border guards).
- Insurgent groups (movements that primarily rely on subversion and violence to change the status quo).
- Guerrilla units (irregular indigenous forces operating in occupied territory).
- Criminal organizations (such as gangs, drug cartels, or hackers).

NATION-STATE ACTORS

2-5. Nation-states fall into four basic categories according to their roles in the international community. The categories are core states, transition states, rogue states, and failed or failing states. Countries can move from one category to another, as conditions change.

2-6. The category of *core states* includes more than half of the nearly 200 countries in the world today. These are basically democratic (although to varying degrees) and share common values and interests. Within this larger group, there is an "inner core" of major powers. These are the advanced countries, including the United States, that generally dominate world politics. Most conflict with global consequences will involve the core states in some fashion or another.

2-7. *Transition states* are other larger, industrial-based countries, mostly emerging regional powers, that have the potential to become accepted among the core states, perhaps as major powers. High-end transition states are moving from an industrial-based society to an information-based society. Low-end transition states are seeking to move from an agricultural-based society to an industrial base. As states try to make this transition, there are cycles of political stability and instability, and the outcome of the transition is uncertain. Some transition states may successfully join the ranks of core states and even become major powers within that context. Others may become competitors.

2-8. *Rogue states* are those that are hostile to their neighbors or to core states' interests. These countries attack or threaten to attack their neighbors. They may sell or give armaments to other countries or non-state actors within or outside their region, thus threatening regional or international stability. They can sponsor international terrorism or even confront U.S. military forces operating in the region.

2-9. *Failed or failing states* are fragmented in such a way that a rule of law is absent. Their instability is a threat to their neighbors and the core states. The government has ceased to meet the needs of all its people, and at least parts of the country may have become virtually ungovernable. Entities other than the legitimate government institutions—such as large criminal organizations—may have filled the power vacuum and taken control. The real threat to U.S. forces may come from elements other than the military. In some cases, the government might be able to control the population and meet the people's needs, but only with outside support—perhaps from countries or groups opposed to U.S. interests. Failed or failing states often harbor groups antagonistic to the United States and its interests.

NON-STATE ACTORS

2-10. Non-state actors are those that do not represent the forces of a particular nation-state. Such non-state elements include rogue actors as well as third-party actors.

2-11. Like rogue states, *rogue actors* are hostile to other actors. However, they may be present in one country or extend across several countries. Examples include insurgents, guerrillas, mercenaries, and transnational or subnational political movements. Particular sources of danger are terrorists and drug-trafficking or criminal organizations, since they may have the best technology, equipment, and weapons available, simply because they have the money to buy them. These non-state rogue actors may use terror tactics and militarily unconventional methods to achieve their goals.

2-12. *Third-party actors* may not be hostile to other actors. However, their presence, activities, and interests can affect the ability of military forces to accomplish their mission. These third-party actors can include—
- Refugees and internally displaced persons.
- International humanitarian relief agencies.
- Transnational corporations.
- News media.

These individuals and groups bring multiple sources of motivation, ideology, interests, beliefs, or political affiliations into consideration. They may be sources of civil unrest. Their presence may require military forces to consider the potential impacts of traffic congestion, demonstrations, sabotage, and information manipulation.

REGULAR MILITARY FORCES

2-13. Regular military forces are the regulated armed forces of a state or alliance of states with the specified function of military offensive and defensive capabilities in legitimate service to the state or alliance. Traditional capabilities of regular military forces normally are intended to accomplish one or more of the following objectives:
- Defeat an adversary's armed forces.
- Destroy an adversary's war-making capacity.
- Seize or retain territory.

These descriptors are consistent with the U.S. DOD definition of traditional warfare per DOD Directive (DODD) 3000.7.

2-14. Other legitimate functions of regular military forces can include a wide range of stability and support missions in concert with state policies and programs. These can include national disaster response, or assistance to province or district government to counter lawlessness, riot, or insurrection.

IRREGULAR FORCES

2-15. *Irregular forces* are armed individuals or groups who are not members of the regular armed forces, police, or other internal security forces (JP 3-24). *Irregular warfare* is a violent struggle among state and non-state actors for legitimacy and influence over the relevant population(s) (JP 1).

REGULAR VERSUS IRREGULAR FORCES

2-16. *Traditional warfare* is a form of warfare between the regulated militaries of nation-states, or alliances of states (DODD 3000.7). In contrast, *unconventional warfare* encompasses a broad spectrum of military and paramilitary operations which are normally of long duration and usually conducted through, with, or by indigenous or surrogate forces (JP 3-05). Traditional armed forces characterize standing military units of a nation-state. A nation-state may also have capabilities such as border guard units, constabulary, and law enforcement organizations that may have an assigned paramilitary role. Irregular forces can exhibit a mixed capability of insurgent, guerrilla, and armed criminal elements. Traditional military units may also be involved directly or indirectly in coordination with irregular warfare operations.

2-17. Irregular forces favor indirect and asymmetric approaches. These approaches may employ the full range of military and other capacities, in order to erode an opponent's power, influence, and will. Irregular warfare typically involves a protracted conflict that involves state and non-state forces in a regional area. However, such a conflict can be readily connected to transnational actions due to globalization on political, economic, and financial fronts.

2-18. Different types of irregular forces may use varied levels of violence or nonviolence to exert influence. Access to technology can impact irregular force operations. Some forces may use low-technology approaches to counter the capabilities of a superpower. Yet, a constant search for improved technologies will parallel a constantly changing set of operational conditions.

2-19. The actions of irregular forces are not a lesser form of conflict below the threshold of warfare. At the tactical level, they can apply tactics, techniques, and procedures common to regular forces but do so with asymmetric applications and means. However, irregular forces also can use methods such as guerrilla warfare, terrorism, sabotage, subversion, coercion, and criminal activities.

2-20. Adversaries faced with the conventional warfighting capacity of the U.S. Army and joint or combined forces partners are likely to choose to fight using a hybrid of traditional, irregular, and/or criminal capabilities as a way to achieve their strategic objectives. A strategy of U.S. adversaries will be to degrade and exhaust U.S. forces rather than cause a direct U.S. military defeat.

2-21. The definition of irregular warfare highlights a key issue of a relevant population and the intention to damage an opponent's influence over that population. The population can be defined in many aspects and may describe itself in terms such as its culture, ethnicity, familial lineage, theology, ideology, or geographic locale. When confronting the United States, regular or irregular forces will seek to undermine and erode the national power, influence, and will of the United States and any strategic partners to exercise political authority over a relevant population.

DEFINING ENEMY COMBATANTS

2-22. The DOD defines an *enemy combatant* as "in general, a person engaged in hostilities against the United States or its coalition partners during an armed conflict" (JP 1-02 from DODD 2311.01E). Other essential terms are *lawful enemy combatant* and *unlawful enemy combatant*. Definitions are provided in figure 2-1 on page 2-4.

Chapter 2

> **Enemy Combatant**
> In general, a person engaged in hostilities against the United States or its coalition partners during an armed conflict. The term enemy combatant includes both "lawful enemy combatants" and "unlawful enemy combatants."
> (DODD 2310.01E)
>
> **Lawful Enemy Combatant**
> Lawful enemy combatants, who are entitled to protections under the Geneva Conventions, include members of the regular armed forces of a State party to the conflict; militia, volunteer corps, and organized resistance movements belonging to a State party to the conflict, which are under responsible command, wear a fixed distinctive sign recognizable at a distance, carry their arms openly, and abide by the laws of war; and members of regular armed forces who profess allegiance to a government or an authority not recognized by the detaining power.
> (DODD 2310.01E)
>
> **Unlawful Enemy Combatant**
> Unlawful enemy combatants are persons not entitled to combat immunity, who engage in acts against the United States or its coalition partners in violation of the laws and customs of war during an armed conflict. ... [The] term *unlawful enemy combatant* is defined to include, but is not limited to, an individual who is or was part of or supporting ... forces that are engaged in hostilities against the United States or its coalition partners.
> (DODD 2310.01E)

Figure 2-1. Combatant definitions

2-23. Combatants can be casually and incorrectly categorized without appropriate attention to what a particular term defines as the purpose, intent, or character of an enemy combatant. Several terms that can easily be misused include paramilitary forces, insurgents, guerrillas, terrorists, militia, and mercenaries. Figure 2-2 provides DOD definitions of the first four terms.

> **Paramilitary Forces**
> [Member of]...forces or groups distinct from the regular armed forces of any country, but resembling them in organization, equipment, training, or mission.
> (JP 3-24)
>
> **Insurgent**
> [Actor in]...organized use of subversion and violence by a group or movement that seeks to overthrow or force change of a governing authority.
> (*Irregular Warfare: Countering Irregular Threats Joint Operating Concept, Version 2.0*)
>
> **Guerrilla**
> A combat participant in guerrilla warfare...[a member of military and paramilitary operations conducted in enemy-held or hostile territory by irregular, predominantly indigenous forces.]
> (JP 3.05.1)]
>
> **Terrorist**
> An individual who commits an act or acts of violence or threatens violence in pursuit of political, religious, or ideological objectives.
> (JP 3-07.2)

Figure 2-2. Paramilitary definitions

Hybrid Threat Components

PARAMILITARY

2-24. *Paramilitary forces* are "forces or groups distinct from the regular armed forces of any country, but resembling them in organization, equipment, training, or mission" (JP 3-24). Thus, there are various types of non-state paramilitary forces, such as insurgents, guerrillas, terrorist groups, and mercenaries. However, there are also nation-state paramilitary forces such as internal security forces, border guards, and police, which are specifically not a part of the regular armed forces of the country.

> *Note.* The term *militia* has acquired many definitions based on the situational context. This context may be the culture; historical traditions such as which group of people have familial, social, theological, or political power; and the external or self-descriptions such forces use in media affairs or propaganda. A generic definition of a militia can parallel the definition of a paramilitary force. However, a nation-state can also have militia that are considered an extension of its armed forces.

INSURGENT

2-25. An *insurgency* is "the organized use of subversion and violence by a group or movement that seeks to overthrow or force change of a governing authority" (JP 3-24). Insurgent organizations have no regular table of organization and equipment structure. The mission, environment, geographic factors, and many other variables determine the configuration and composition of each insurgent organization and its subordinate cells. A higher insurgent organization can include organizations at regional, provincial, district, national, or transnational levels. Higher insurgent organizations can contain a mix of local insurgent and guerrilla organizations. Each of these organizations provides differing capabilities.

2-26. As an insurgent organization develops and grows, it often forms a political headquarters to communicate with the indigenous population, external supporters, and its enemies. The leaders in this central political headquarters direct the insurgency's paramilitary forces and ensure that the insurgency remains focused on reaching its long-term political goals.

GUERRILLA

2-27. A *guerrilla* is "a combat participant in guerrilla warfare" (JP 1-02). *Guerrilla warfare* is "military and paramilitary operations conducted in enemy-held or hostile territory by irregular, predominantly indigenous forces" (JP 3-05.1). A prime characteristic of guerrilla operations is to attack points of enemy weakness and in conditions developed or selected by the guerrilla force. Deception and mobility are critical to achieving surprise and avoiding engagements unless the tactical opportunity weighs heavily in the favor of the guerrilla. At the tactical level, attacks are planned and conducted as sudden, violent, decentralized actions. Principles of rapid dispersion and rapid concentration facilitate these types of operation.

> **The Maoist Example**
>
> Mao Tse-tung's concept of guerrilla war offers a prime example of different types of forces working toward a common purpose, and exemplifies an adaptive concept of waging war. In the China of the 1930s and 1940s, Mao realized that guerrilla warfare was only one of several approaches in mobilizing a population against an enemy. For the means available to the Communist Chinese at the time, they optimized military and political capabilities in localized home guards or militia, as well as guerrilla groups from squad to regiment in size. They also configured regular military units as resources, training, and events allowed such evolution and cooperative forms of combat power.
>
> According to Mao, guerrilla forces evolve gradually from rudimentary paramilitary elements to more traditional military-like forces that plan and operate in conjunction with regular army units of the revolutionary army. Guerrilla warfare cannot be isolated

> in concept or practice from the offensive and defensive actions of the regular army. Trying to separate guerrilla warfare from traditional warfare denies the contributions that each type of force provides to synergy in operational effectiveness. Mao identifies three types of cooperation among guerrillas and orthodox military units: strategic, tactical, and battle cooperation.
>
> Such an organization can optimize conceptual aspects of warfare and is "not dependent for success on the efficient operation of complex mechanical devices, highly organized logistical systems, or the accuracy of electronic computers. It can be conducted in any terrain.... Its basic element is man ... endowed with intelligence, emotions, and will ... politically educated and thoroughly aware of the issues at stake." (*Mao Tse-tung on Guerrilla Warfare*, USMC FMFRP 12-18)
>
> Capabilities or limitations of a hybrid force emerge as conditions change in an OE. The Mao model lists multiple sources for capability that include—
> - Volunteers from the general population.
> - Regular unit soldiers detailed temporarily to a guerrilla force.
> - Regular unit soldiers detailed permanently to a guerrilla force as a cadre.
> - Combinations of regular unit members and locally recruited civilians.
> - Local militia or self defense home guard members.
> - Deserters from the enemy forces.
> - Even former "bandits and bandit groups."

TERRORIST

2-28. A *terrorist* is "an individual who commits an act or acts of violence or threatens violence in pursuit of political, religious, or ideological objectives" (JP 3-07.2). A *terrorist group* is "any number of terrorists who assemble together, have a unifying relationship, or are organized for the purpose of committing an act or acts of violence or threatens violence in pursuit of their political, religious, or ideological objectives" (JP 3-07.2). Categorizing terrorist groups by their affiliation with governments or supporting organizations can provide insight in terrorist intent and capability. Terrorist groups can align as state-directed, state-sponsored, or non-state supported organizations. In some cases, the state itself can be a terrorist regime.

MERCENARY

2-29. *Mercenaries* are armed individuals who use conflict as a professional trade and service for private gain. Those who fall within that definition are not considered combatants. However, those who take direct part in hostilities can be considered unlawful enemy combatants. The term *mercenary* applies to those acting individually and in formed units. Soldiers serving officially in foreign armed forces are not mercenaries. Loan service personnel sent to help train the soldiers of other countries as part of an official training agreement between sovereign governments are not mercenaries even if they take a direct part in hostilities.

2-30. In accordance with the Geneva Conventions, mercenaries are individuals who act individually or act as a member of a formed group, and have all the following characteristics:
- Are recruited locally or abroad in order to fight in an armed conflict.
- Are operating directly in the hostilities.
- Are motivated by the desire for private gain. (They are promised, by or on behalf of a party to the conflict, material compensation substantially in excess of that promised or paid to the combatants of similar rank and functions in the armed forces of that party.)
- Are neither nationals of a party to the conflict nor residents of territory controlled by a party to the conflict.
- Are not members of the armed forces of a party to the conflict.
- Are not on official military duty representing a country that is not involved in the conflict such as a legitimate loan service or training appointment.

Hybrid Threat Components

CRIMINAL ORGANIZATIONS

2-31. There is no part of the world that is criminal-free. Therefore, there will always be criminal elements present in any OE. The only question is whether those criminal organizations will find it in their interests to become part of a hybrid threat and to perform some of the functions required to achieve common goals and objectives.

2-32. Criminal organizations are normally independent of nation-state control. However, large-scale criminal organizations often extend beyond national boundaries to operate regionally or worldwide and include a political influence component. Individual criminals or small gangs do not normally have the capability to adversely affect legitimate political, military, and judicial organizations. However, large-scale criminal organizations can challenge governmental authority with capabilities and characteristics similar to a paramilitary force.

2-33. By mutual agreement or when their interests coincide, criminal organizations may become affiliated with other actors such as insurgents or individuals. They may provide capabilities similar to a private army for hire. Insurgents or guerrillas controlling or operating in the same area as a criminal organization can provide security and protection to the criminal organization's activities in exchange for financial assistance, intelligence, arms and materiel, or general logistical support. On behalf of the criminal organization, guerrilla or insurgent organizations can—

- Create diversionary actions.
- Conduct reconnaissance and early warning.
- Conduct money laundering, smuggling, or transportation.
- Conduct civic actions.

Their mutual interests can include preventing U.S. or government forces from interfering in their respective activities.

2-34. Some criminals may form loosely affiliated organizations that have no true formal structure. Nevertheless, even low-capability criminals sometimes can impact events through opportunistic actions. Criminal violence degrades a social and political environment. As small criminal organizations expand their activities to compete with or support of long-established criminal organizations, criminals may seek neutralize or control political authority in order to improve their ability to operate successfully and discourage rival criminal enterprises.

2-35. At times, criminal organizations might also be affiliated with nation-state military or paramilitary actors. In time of armed conflict or support to a regional insurgency, a state can encourage and materially support criminal organizations to commit actions that contribute to the breakdown of civil control in a neighboring country.

USE OF WMD

2-36. The intent of hybrid threats to obtain and use weapons of mass destruction (WMD) is one of the most serious contemporary threats to regional neighbors or even the U.S. homeland. The means of attack can range from a highly sophisticated weapon system such as a nuclear bomb to a rudimentary improvised radiological device. The specter of chemical contamination or biological infection adds to the array of weapons. Although high-yield explosives have not been traditionally recognized as WMD, high-yield and some low-yield explosives have caused significant devastating effects on people and places. With any type of WMD, the hybrid threat's desired outcome could involve any or all of the following:

- Producing mass casualties.
- Massive damage of physical infrastructure and/or the economy.
- Extensive disruption of activities and lifestyles.

2-37. The threat of WMD use is present across the entire spectrum of conflict. Potential exists for WMD use with individual acts of wanton damage or destruction of property or person, as well as operations conducted by organized violent groups or rogue states. Hybrid threats may include organizations with demonstrated global reach capabilities and the intention to acquire and use WMD. For example, an international terrorist network with WMD capability could target extragegional military forces either in their homeland or as they are deploying into the region. It would patiently await the opportunity to achieve maximum operational or strategic impact of its use of WMD.

2-38. Ultimately, a significant impact on a large population would be an intimidating psychological effect from physical and emotional stress. Simply stated, the potential for mass injury or death, as well as mass damage or destruction, presents a compelling requirement for protective measures and increased assurance to counter public harm, anxiety, and fear.

2-39. Three general trends with impact on hybrid threat use of WMD are micro-actors, sophistication, and overlap with transnational crime. Each of these trends can pose a critical danger by linking intent with WMD capability. Growing numbers of small independent actors can manipulate advanced technologies to gain knowledge and means while masking their operational or tactical plans. Sophistication involves a combination of global information systems, financial resources, and practical exchange of ideas. Transnational criminals demonstrate themselves to be a valuable network to assist other hybrid threat components with enhanced mobility, improved support, and concealed actions.

PART TWO

The Hybrid Threat for Training

Part two of this TC focuses on the Hybrid Threat (HT) for U.S. Army training. The HT is a realistic and relevant composite of actual hybrid threats. This composite constitutes the enemy, adversary, or threat whose military and/or paramilitary forces are represented as an opposing force (OPFOR) in training exercises. The following chapters will discuss the strategy, operations, tactics, and organizations of the HT. For more detail, the reader should consult other TCs in the 7-100 series and supporting products.

The OPFOR, when representing a hybrid threat, must be a challenging, uncooperative adversary or enemy. It must capable of stressing any or all warfighting functions and mission-essential tasks of the U.S. armed force being trained. Training for the challenges of contemporary operational environments requires an OPFOR that is "a plausible, flexible military and/or paramilitary force representing a composite of varying capabilities of actual worldwide forces, used in lieu of a specific threat force, for training and developing U.S. forces" (Army Regulation 350-2).

The commander of a U.S. unit plans and conducts training based on the unit's mission essential task list and priorities of effort. The commander establishes the conditions in which to conduct training to standards. These conditions should include an OPFOR that realistically challenges the ability of the U.S. unit to accomplish its tasks. Training requirements will determine whether the OPFOR's capabilities are fundamental, sophisticated, or a combination of these.

As real-world conditions and capabilities change, elements of OPFOR doctrine, organization, and equipment capabilities will evolve also. The OPFOR will remain capable of presenting realistic and relevant challenges that are appropriate to meet evolving training requirements. Military forces may have paramilitary forces acting in loose affiliation with them, or acting separately from them within the same training environment. These relationships depend on the scenario, which is crafted based on the training requirements and conditions of the Army unit being trained.

Chapter 3

Hybrid Threat Strategy

Hybrid Threat (HT) strategy is sophisticated, comprehensive, and multi-dimensional. In pursuit of its strategic goals, the HT is prepared to conduct four basic types of strategic-level courses of action (COAs). It calls these strategic operations, regional operations, transition operations, and adaptive operations. Each COA involves the use of all four instruments of power (not just military and/or paramilitary, but also diplomatic-political, informational, and economic means), but to different degrees and in

Chapter 3

different ways. The strategic operations COA overarches the other three, which also serve as basic operational designs (see chapter 4).

STRATEGIC OPERATIONS

3-1. What the HT calls "strategic operations" is a COA that uses all instruments of power in peace and war to achieve the HT's goals by attacking the enemy's strategic centers of gravity. It is a universal strategic COA the HT would use to deal with all situations, against all kinds of opponents, potential opponents, or neutral parties.

Note. The introduction and chapters 1 and 2 use the terms *enemy* and *adversary* to refer to various nation-state or non-state actors that threaten or oppose U.S. interests. However, the remainder of this TC will use *enemy* and *adversary* to refer to an enemy or adversary of the actors who make up the HT. Likewise, *friendly* refers to the HT.

3-2. Strategic operations are a continuous process not limited to wartime or preparation for war. Once war begins, they continue during regional, transition, and adaptive operations and complement those operations. Each of the latter three types of operations occurs only during war and only under certain conditions. Transition operations can overlap regional and adaptive operations. See figure 3-1.

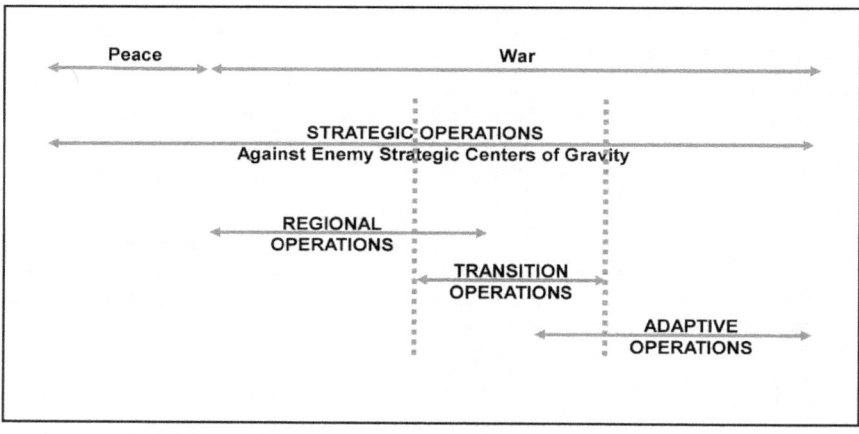

Figure 3-1. Strategic operations and other courses of action

STRATEGIC GOALS AND OPERATIONAL DESIGNS

3-3. The first aim of strategic operations is to preclude an extraregional power (a foreign power, such as the United States, from outside the region) from intervening in the HT's region. (See the section on Strategic Preclusion later in this chapter.) If preclusion is not possible, the aim becomes that of getting the extraregional force to leave before it can achieve the goals of its intervention. Strategic operations may make use of operational designs unnecessary, or they may continue while the HT operates along one of its operational designs.

3-4. HT strategy is designed to achieve one or more specific strategic goals within the HT's region. Therefore, it typically starts with actions directed at an opponent within the region.

Hybrid Threat Strategy

3-5. If possible, the HT will attempt to achieve its ends without resorting to armed conflict. Accordingly strategic operations are not limited to military and/or paramilitary means and usually do not begin with armed conflict. The HT may be able to achieve the desired goal through pressure applied by other than military or paramilitary instruments of power, perhaps with the mere threat of using that power against the regional opponent. These actions would fall under the general framework of "strategic operations."

3-6. When nonmilitary means are not sufficient or expedient, the HT may resort to armed conflict as a means of creating conditions that lead to the desired end state. However, strategic operations continue even if a particular regional threat or opportunity causes the HT to undertake "regional operations" that include military means.

3-7. Prior to initiating armed conflict and throughout the course of armed conflict with its regional opponent, the HT continues to conduct strategic operations to preclude intervention by outside actors. Such actors could include other regional neighbors or an extraregional power that could overmatch the HT's forces. However, plans for those regional operations always include branches and sequels for dealing with the possibility of intervention by an extraregional power. When unable to limit the conflict to regional operations, the HT is prepared to engage extraregional forces through "transition and adaptive operations." Usually, the HT does not shift directly from regional to adaptive operations. The transition is incremental and does not occur at a single, easily identifiable point. If the HT perceives that intervention is likely, transition operations may begin simultaneously with regional and strategic operations.

3-8. Transition operations overlap both regional and adaptive operations. Transition operations allow the HT to shift to adaptive operations or back to regional operations. At some point, the HT either seizes an opportunity to return to regional operations, or it reaches a point where it must complete the shift to adaptive operations. Even after shifting to adaptive operations, the HT tries to set conditions for transitioning back to regional operations.

3-9. If an extraregional power were to have significant forces already deployed in the region prior to the outbreak of hostilities, the HT would not be able to conduct regional operations using a typical conventional design without first neutralizing those forces. In this case, the HT would first use strategic operations, with all means available, to put pressure on the already present extraregional force to withdraw from the region or at least remain neutral in the regional conflict. Barring that, strategic operations could still aim at keeping the extraregional power from committing additional forces to the region and preventing his forces already there from being able to fully exercise their capabilities. If the extraregional force is still able to intervene, the HT would have to start with adaptive operations.

3-10. Eventually, the HT would seek to move back into transition operations. If it could neutralize or eliminate the extraregional force, it could finally complete the transition to regional operations and thus achieve its strategic goals.

MEANS

3-11. Strategic operations apply all four instruments of power, in varying combinations depending on the conditions. In most cases, the diplomatic-political, informational, and economic means tend to dominate. During strategic operations, military and/or paramilitary means are most often used to complement those other instruments of power to achieve HT goals. For example, the military and/or paramilitary means are likely to be used against key political or economic centers or tangible targets whose destruction affects intangible centers of gravity, rather than against military targets for purely military objectives.

Note. One of the four instruments of power available to the HT is called "diplomatic-political" because non-state components of the HT would not have diplomatic means. However, they could exert political pressure or influence.

3-12. Against such targets, the HT will employ all means available, including—
- Diplomatic initiatives.
- Political influence.

Chapter 3

- Information warfare (INFOWAR).
- Economic pressure.
- Terrorist attacks.
- HT-sponsored insurgency.
- Direct action by special-purpose forces (SPF).
- Long-range precision fires.
- Even weapons of mass destruction (WMD) against selected targets.

These efforts often place noncombatants at risk and aim to apply diplomatic-political, economic, and psychological pressure by allowing the enemy no sanctuary.

3-13. The use of diplomatic-political or economic means or pressure is always orchestrated at the highest level, as is strategic INFOWAR. Even with the military instrument of power, actions considered part of strategic operations require a conscious, calculated decision and direction or authorization by the HT leadership, which is not entirely military in its makeup.

Targets

3-14. Strategic operations target the enemy's strategic centers of gravity. They attack the intangible components of the enemy's efforts against the HT. They primarily target those elements that can most affect factors such as—

- Enemy soldiers' and leaders' confidence.
- Political and diplomatic decisions.
- Public opinion.
- The interests of private institutions.
- National will and the collective will and commitment of alliances and coalitions.

National will is not just the will to fight, but also the will to intervene by other than military means.

3-15. It may not be readily apparent to outside parties whether specific actions by the HT's various instruments of power are part of strategic operations or part of another strategic-level COA occurring simultaneously. In fact, one action could conceivably fulfill both purposes. For example, a demoralizing defeat that could affect the enemy's strategic centers of gravity could also be a defeat from an operational or tactical viewpoint. In other cases, a particular action on the battlefield might not make sense from a tactical or operational viewpoint, but could achieve a strategic purpose. Its purpose may be to inflict mass casualties or destroy high-visibility enemy systems in order to weaken the enemy's national will to continue the intervention.

3-16. Likewise, victims of terror tactics may not be able to tell whether they were attacked by actual terrorists (independent or affiliated) or by insurgents, criminal organizations, or SPF using terror tactics. However, the results are the same. From the HT's point of view, it can exploit the effects such attacks have on the enemy's tangible capabilities and/or his intangible centers of gravity. In makes no difference whether the HT planned and carried out the attack or was merely able to capitalize on it and reap the benefits of someone else's action. Even when the HT is responsible, there is opportunity for plausible deniability.

Timeframe

3-17. Strategic operations occur continuously, from prior to the outbreak of war to the post-war period. They can precede war, with the aim of deterring other regional actors from actions hostile to the HT's interests or compelling such actors to yield to the HT's will.

3-18. The HT is always applying its diplomatic-political, informational, and economic instruments of power. Even in peacetime, the very presence of the HT's military and/or paramilitary power gives the HT leverage and influence in regional affairs. Another tool for expanding the HT's influence is the use of peacetime programs and training exercises that regular armed forces that are part of the HT conduct to—

- Shape the international environment.
- Open communications and improve mutual understanding with other countries.

Hybrid Threat Strategy

- Improve interoperability with allies and potential allies.

The HT can also foster military or economic cooperation based on historical relationships. Thus, it may be possible for the HT to achieve its strategic goals without ever resorting to armed conflict.

3-19. In wartime, strategic operations become an important, powerful component of the HT's strategy for total war. They occur concurrently with regional, transition, and adaptive operations and can change the course of other strategic-level COAs or even bring the war to an end. Strategic operations may continue even after termination of the armed conflict. If the HT succeeds in defeating the extraregional force or at least forces it to withdraw from the region, this victory enhances the HT's status both regionally and globally. It will take advantage of this status to pursue its strategic goals. Should the HT lose this war as judged from conventional political or military standards, but still survive as a nation or non-state entity, it may be able to claim victory.

STRATEGIC INFORMATION WARFARE

3-20. An important component the HT's strategy for total war is the conduct of *information warfare* (INFOWAR), which the HT defines as specifically planned and integrated actions taken to achieve an information advantage at critical points and times. The goal is to influence an enemy's decisionmaking through his collected and available information, information systems, and information-based processes, while retaining the HT's ability to employ the same. In the context of total war, INFOWAR encompasses all instruments of power. It is not just a military function and concept. Thus, the HT applies INFOWAR at every level of conflict and in peacetime interactions with other actors.

3-21. Despite the fact that the HT refers to it as "warfare," INFOWAR exists in peacetime as well as during war. In peacetime, INFOWAR involves struggle and competition, rather than actual "warfare," as states and non-state actors maneuver and posture to protect their own interests, gain an advantage, or influence others.

3-22. During times of crisis and war, INFOWAR activities continue and intensify. Defensive INFOWAR measures are more strictly enforced, while some of the more offensive elements of INFOWAR come to the fore. Even the subtler elements may become more aggressive and assertive.

THE STRATEGIC DIMENSION

3-23. Because of its significance to the overall achievement of the HT's strategy, INFOWAR at the strategic level receives special attention. Strategic INFOWAR is the synergistic effort of the HT to control or manipulate information events, be they diplomatic, political, economic, or military in nature. Specifically, the HT defines *strategic INFOWAR* as any attack (digital, physical, or cognitive) against the information base of an adversary nation's critical infrastructures.

3-24. The ultimate goal of strategic INFOWAR is strategic disruption and damage to the overall strength of an opponent. This disruption also focuses on the shaping of foreign decisionmakers' actions to support the HT's strategic objectives and goals. Perception management activities are critical to strategic INFOWAR. The HT attempts to use all forms of persuasion and global media to win the "battle of the story."

3-25. Strategic INFOWAR can undermine an extraregional power's traditional advantage of geographic sanctuary from strategic attack. Strategic INFOWAR is not confined to a simple zone of territory, but can extend globally to encompass attacks on an opponent's homeland or the homelands of various military coalition members.

3-26. In addition to using all its own assets, the HT will seek third-party actors or outside resources to support its overall information strategy. The HT facilitates these shadow networks as necessary and continuously cultivates and maintains them during peacetime.

Chapter 3

ELEMENTS OF INFOWAR

3-27. Integrated within INFOWAR doctrine are the following elements:
- **Electronic warfare (EW).** Measures conducted to control or deny the enemy's use of the electromagnetic spectrum, while ensuring its use by the HT.
- **Deception.** Measures designed to mislead the enemy by manipulation, distortion, or falsification of information to induce him to act in a manner prejudicial to his interests.
- **Physical destruction.** Measures to destroy critical components of the enemy's information infrastructure.
- **Protection and security measures.** Measures to protect the HT's information infrastructure and to deny protected information to other actors.
- **Perception management.** Measures aimed at creating a perception of truth that best suits HT objectives. Perception management uses a combination of true, false, and misleading information targeted at the local populace and/or external actors. This element is crucial to successful strategic INFOWAR. The HT is continuously looking for ways to sway international opinion in its favor or impact critical foreign strategic decisionmakers.
- **Information attack (IA).** Measures focused on the intentional disruption of digital information in a manner that supports a comprehensive strategic INFOWAR campaign. IAs focus exclusively on the manipulation or degradation of the information moving throughout the information environment. Unlike computer warfare attacks that target the information systems, IAs target the information itself.
- **Computer warfare.** Measures ranging from unauthorized access (hacking) of information systems for intelligence collection purposes to the insertion of destructive viruses and deceptive information into enemy computer systems. Such attacks focus on the denial of service and/or disruption or manipulation of the infrastructure's integrity. Strategic INFOWAR typically targets critical nodes or hubs, rather than targeting the entire network or infrastructure.

3-28. The seven elements of INFOWAR do not exist in isolation from one another and are not mutually exclusive. The overlapping of functions, means, and targets makes it necessary that they all be integrated into a single INFOWAR plan. However, effective execution of strategic INFOWAR does not necessary involve the use of all elements concurrently. In some cases, one element may be all that is required to successfully execute a strategic INFOWAR action or a supporting action at the operational or tactical level. The use of each element or a combination of elements is determined by the overall situation and specific strategic goals.

STRATEGIC PRECLUSION

3-29. Strategic preclusion seeks to completely deter extraregional involvement or severely limit its scope and intensity. The HT would attempt to achieve strategic preclusion in order to reduce the influence of the extraregional power or to improve its own regional or international standing. It would employ all its instruments of power to preclude direct involvement by the extraregional power. Actions can take many forms and often contain several lines of operation working simultaneously.

3-30. The primary target of strategic preclusion is the extraregional power's national will. First, the HT would conduct diplomatic-political and perception management activities aimed at influencing regional, transnational, and world opinion. For example, the HT might use a disinformation campaign to discredit the legitimacy of diplomatic or economic sanctions imposed upon it. The extraregional power's economy and military would be secondary targets, with both practical and symbolic goals. This might include using global markets and international financial systems to disrupt the economy of the extraregional power, or conducting physical and information attacks against critical economic centers. Similarly, the military could be attacked indirectly by disrupting its power projection, mobilization, and training capacity. Preclusive actions are likely to increase in intensity and scope as the extraregional power moves closer to military action. If strategic preclusion fails, the HT will turn to operational methods that attempt to limit the scope of extraregional involvement or cause it to terminate quickly.

Chapter 4

Hybrid Threat Operations

Of the four types of strategic-level courses of action outlined in chapter 3, regional, transition, and adaptive operations are also operational designs. This chapter explores those designs and outlines the Hybrid Threat's (HT's) principles of operation against an extraregional power. For more detail, see FM 7-100.1.

OPERATIONAL DESIGNS

4-1. The HT employs three basic operational designs:
- **Regional operations.** Actions against regional adversaries and internal threats.
- **Transition operations.** Actions that bridge the gap between regional and adaptive operations and contain some elements of both. The HT continues to pursue its regional goals while dealing with the development of outside intervention with the potential for overmatching the HT's capabilities.
- **Adaptive operations.** Actions to preserve the HT's power and apply it in adaptive ways against overmatching opponents.

4-2. Each of these operational designs is the aggregation of the effects of tactical, operational, and strategic actions, in conjunction with the other three instruments of power, that contribute to the accomplishment of strategic goals. The type(s) of operations the HT employs at a given time will depend on the types of threats and opportunities present and other conditions in the operational environment (OE). Figure 4-1 illustrates the HT's basic conceptual framework for the three operational designs.

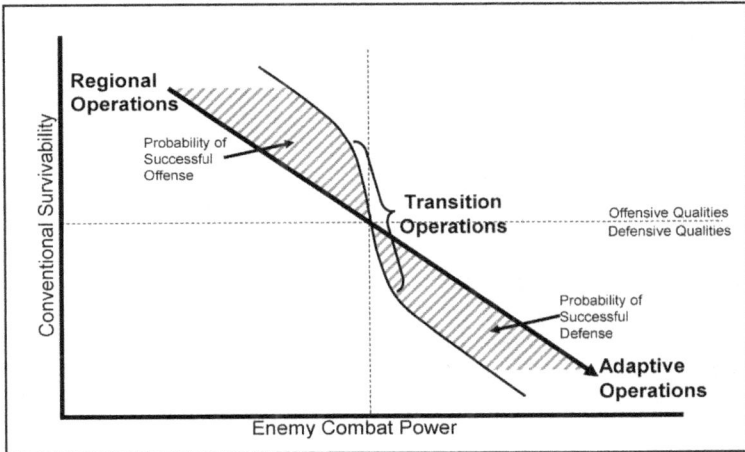

Figure 4-1. Operational designs

Chapter 4

REGIONAL OPERATIONS

4-3. Against opponents from within its region, the HT may conduct "regional operations" with a relatively high probability of success in primarily offensive actions. HT offensive operations are characterized by using all available HT components to saturate the OE with actions designed to disaggregate an opponent's capability, capacity, and will to resist. These actions will not be limited to attacks on military and security forces, but will affect the entire OE. The opponent will be in a fight for survival across many of the variables of the OE: political, military, economic, social, information, and infrastructure.

4-4. HT offensive operations seek to—
- Destabilize control.
- Channel actions of populations.
- Degrade key infrastructure.
- Restrict freedom of maneuver.
- Collapse economic relationships.
- Retain initiative.

These operations paralyze those elements of power the opponent possesses that might interfere with the HT's goals

4-5. The HT may constantly shift which components and sets of components act to affect each variable. For example, regular forces may attack economic targets while criminal elements simultaneously act against an enemy military base or unit in one action, and then in the next action their roles may be reversed. In another example, information warfare (INFOWAR) assets may attack a national news broadcast one day, a military command and control (C2) network the next day, and a religious gathering a day later. In addition to military, economic, and information aspects of the OE, HT operations may include covert and overt political movements to discredit incumbent governments and serve as a catalyst to influence popular opinion for change. The synergy of these actions creates challenges for opponents of the HT in that it is difficult to pinpoint and isolate specific challenges.

4-6. The HT may possess an overmatch in some or all elements of power against regional opponents. It is able to employ that power in an operational design focused on offensive action. A weaker regional neighbor may not actually represent a threat, but rather an opportunity that the HT can exploit. To seize territory or otherwise expand its influence in the region, the HT must destroy a regional enemy's will and capability to continue the fight. It will attempt to achieve strategic decision or achieve specific regional goals as rapidly as possible, in order to preclude regional alliances or outside intervention.

4-7. During regional operations, the HT relies on its continuing strategic operations (see chapter 3) to preclude or control outside intervention. It tries to keep foreign perceptions of its actions during a regional conflict below the threshold that will invite in extraregional forces. The HT wants to achieve its objectives in the regional conflict, but has to be careful how it does so. It works to prevent development of international consensus for intervention and to create doubt among possible participants. Still, at the very outset of regional operations, it lays plans and positions forces to conduct access-limitation operations in the event of outside intervention.

TRANSITION OPERATIONS

4-8. Transition operations serve as a pivotal point between regional and adaptive operations. The transition may go in either direction. The fact that the HT begins transition operations does not necessarily mean that it must complete the transition from regional to adaptive operations (or vice versa). As conditions allow or dictate, the "transition" could end with the HT conducting the same type of operations as before the shift to transition operations.

4-9. The HT conducts transition operations when other regional and/or extraregional forces threaten its ability to continue regional operations in a conventional design against the original regional enemy. At the point of shifting to transition operations, the HT may still have the ability to exert all instruments of power against an overmatched regional enemy. Indeed, it may have already defeated its original adversary.

However, its successful actions in regional operations have prompted either other regional actors or an extraregional actor to contemplate intervention. The HT will use all means necessary to preclude or defeat intervention.

4-10. Although the HT would prefer to achieve its strategic goals through regional operations, it has the flexibility to change and adapt if required. Since the HT assumes the possibility of extraregional intervention, its plans will already contain thorough plans for transition operations, as well as adaptive operations, if necessary.

4-11. When an extraregional force starts to deploy into the region, the balance of power begins to shift away from the HT. Although the HT may not yet be overmatched, it faces a developing threat it will not be able to handle with normal, "conventional" patterns of operation designed for regional conflict. Therefore, the HT must begin to adapt its operations to the changing threat.

4-12. While the HT is in the condition of transition operations, an operational- or tactical-level commander will still receive a mission statement in plans and orders from his higher authority stating the purpose of his actions. To accomplish that purpose and mission, he will use as much as he can of the conventional patterns of operation that were available to him during regional operations and as much as he has to of the more adaptive-type approaches dictated by the presence of an extraregional force.

4-13. Even extraregional forces may be vulnerable to "conventional" operations during the time they require to build combat power and create support at home for their intervention. Against an extraregional force that either could not fully deploy or has been successfully separated into isolated elements, the HT may still be able to use some of the more conventional patterns of operation. The HT will not shy away from the use of military means against an advanced extraregional opponent so long as the risk is commensurate with potential gains.

4-14. Transition operations serve as a means for the HT to retain the initiative and pursue its overall strategic goals. From the outset, one of the HT's strategic goals would have been to defeat any outside intervention or prevent it from fully materializing. As the HT begins transition operations, its immediate goal is preservation of its instruments of power while seeking to set conditions that will allow it to transition back to regional operations. Transition operations feature a mixture of offensive and defensive actions that help the HT control the tempo while changing the nature of conflict to something for which the intervening force is unprepared. Transition operations can also buy time for the HT's strategic operations to succeed.

4-15. There are two possible outcomes to transition operations. If the extraregional force suffers sufficient losses or for other reasons must withdraw from the region, the HT's operations may begin to transition back to regional operations, again becoming primarily offensive. If the extraregional force is not compelled to withdraw and continues to build up power in the region, the HT's transition operations may begin to gravitate in the other direction, toward adaptive operations.

ADAPTIVE OPERATIONS

4-16. Generally, the HT conducts adaptive operations as a consequence of intervention from outside the region. Once an extraregional force intervenes with sufficient power to overmatch the HT, the full conventional design used in regionally-focused operations is no longer sufficient to deal with this threat. The HT has developed its techniques, organization, capabilities, and strategy with an eye toward dealing with both regional and extraregional opponents. It has already planned how it will adapt to this new and changing threat and has included this adaptability in its methods.

4-17. The HT's immediate goal is survival. However, its long-term goal is still the expansion of influence. In the HT's view, this goal is only temporarily thwarted by the extraregional intervention. Accordingly, planning for adaptive operations focuses on effects over time. The HT believes that patience is its ally and an enemy of the extraregional force and its intervention in regional affairs.

4-18. The HT believes that adaptive operations can lead to several possible outcomes. If the results do not completely resolve the conflict in its favor, they may at least allow it to return to regional operations. Even a stalemate may be a victory, as long as it preserves enough of its instruments of power and lives to fight another day.

4-19. When an extraregional power intervenes, the HT has to adapt its patterns of operation. It still has the same forces and technology that were available to it for regional operations, but must use them in creative and adaptive ways. It has already thought through how it will adapt to this new or changing threat in general terms (See Principles of Operation versus an Extraregional Power, below.) It has already developed appropriate branches and sequels to its core plans and does not have to rely on improvisation. During the course of combat, it will make further adaptations, based on experience and opportunity.

4-20. Even with the intervention of an advanced extraregional power, the HT will not cede the initiative. It may employ military means so long as this does not either place its survival at risk or risk depriving it of sufficient force to remain a significant influence in its region after the extraregional intervention is over. The primary objectives are to—
- Preserve power.
- Degrade the enemy's will and capability to fight.
- Gain time for aggressive strategic operations to succeed.

4-21. The HT will seek to conduct adaptive operations in circumstances and terrain that provide opportunities to optimize its own capabilities and degrade those of the enemy. It will employ a force that is optimized for the terrain or for a specific mission. For example, it will use its antitank capability, tied to obstacles and complex terrain, inside a defensive structure designed to absorb the enemy's momentum and fracture his organizational framework.

4-22. The types of adaptive actions that characterize adaptive operations can also serve the HT well in regional or transition operations, at least at the tactical and operational levels. However, once an extraregional force becomes fully involved in the conflict, the HT will conduct adaptive actions more frequently and on a larger scale.

PRINCIPLES OF OPERATION VERSUS AN EXTRAREGIONAL POWER

4-23. The HT assumes the distinct possibility of intervention by a major extraregional power in any regional conflict. It views the United States as the most advanced extraregional force it might have to face. Like many other countries and non-state actors, the HT has studied U.S. military forces and their operations and is pursuing lessons learned based on its assessments and perceptions. The HT is therefore using the United States as its baseline for planning adaptive approaches for dealing with the strengths and weaknesses of an extraregional force. It believes that preparing to deal with intervention by U.S. forces will enable it to deal effectively with those of any other extraregional power. Consequently, it has devised the following principles for applying its various instruments of diplomatic-political, informational, economic, and military power against this type of threat.

ACCESS LIMITATION

4-24. Extraregional enemies capable of achieving overmatch against the HT must first enter the region using power-projection capabilities. Therefore, the HT's force design and investment strategy is focused on access limitation in order to—
- Selectively deny, delay, and disrupt entry of extraregional forces into the region.
- Force them to keep their operating bases beyond continuous operational reach.

This is the easiest manner of preventing the accumulation of enemy combat power in the region and thus defeating a technologically superior enemy.

4-25. Access limitation seeks to affect an extraregional enemy's ability to introduce forces into the theater. Access-limitation operations do not necessarily have to deny the enemy access entirely. A more realistic goal is to limit or interrupt access into the theater in such a way that the HT's forces are capable of dealing with them. By limiting the amount of force or the options for force introduction, the HT can create conditions that place its conventional capabilities on a par with those of an extraregional force. Capability is measured in terms of what the enemy can bring to bear in the theater, rather than what the enemy possesses.

4-26. The HT's goal is to limit the enemy's accumulation of applicable combat power to a level and to locations that do not threaten the accomplishment of the HT's overall strategic goals. This may occur through many methods. For example, the HT may be able to limit or interrupt the enemy's deployment through actions against his aerial and sea ports of debarkation (APODs and SPODs) in the region. Hitting such targets also has political and psychological value. The HT will try to disrupt and isolate enemy forces that are in the region or coming into it, so that it can destroy them piecemeal. It might exploit and manipulate international media to paint foreign intervention in a poor light, decrease international resolve, and affect the force mix and rules of engagement (ROE) of the deploying extraregional forces.

CONTROL TEMPO

4-27. The HT initially employs rapid tempo in an attempt to conclude regional operations before an extraregional force can be introduced. It will also use rapid tempo to set conditions for access-limitation operations before the extraregional force can establish a foothold in the region. Once it has done that, it needs to be able to control the tempo—to ratchet it up or down—as is advantageous to its own operational or tactical plans.

4-28. During the initial phases of an extraregional enemy's entry into the region, the HT's forces may employ a high operational tempo to take advantage of the weaknesses inherent in enemy power projection. (Lightly equipped forces are usually the first to enter the region.) This may take the form of attack by the HT's forces against enemy early-entry forces, linked with diplomatic, economic, and informational efforts to terminate the conflict quickly before main enemy forces can be brought to bear. Thus, the HT may be able to force the enemy to conventional closure, rather than needing to conduct adaptive operations later, when overmatched by the enemy.

4-29. An extraregional enemy normally tries to slow the tempo while it is deploying into the region and to speed it up again once it has built up overwhelming force superiority. The HT's forces will try to increase the tempo when the enemy wants to slow it and to slow the tempo at the time when the enemy wants to accelerate it.

4-30. By their nature, offensive operations tend to control time or tempo. Defensive operations tend to determine space or location. Through a combination of defensive and offensive actions, the HT's adaptive operations seek to control both location and tempo.

4-31. If the HT cannot end the conflict quickly, it may take steps to slow the tempo and prolong the conflict. This can take advantage of enemy lack of commitment over time. The preferred HT tactics during this period would be those means that avoid decisive combat with superior forces. These activities may not be linked to maneuver or ground objectives. Rather, they may be intended instead to inflict mass casualties or destroy flagship systems, both of which will reduce the enemy's will to continue the fight.

CAUSE POLITICALLY UNACCEPTABLE CASUALTIES

4-32. The HT will try to inflict highly visible and embarrassing losses on enemy forces to weaken the enemy's domestic resolve and national will to sustain the deployment or conflict. Modern wealthy nations have shown an apparent lack of commitment over time. They have also demonstrated sensitivity to domestic and world opinion in relation to conflict and seemingly needless casualties. The HT believes it can have a comparative advantage against superior forces because of the collective psyche and will of the HT forces and their leadership to endure hardship or casualties, while the enemy may not be willing to do the same.

4-33. The HT also has the advantage of disproportionate interests. The extraregional force may have limited objectives and only casual interest in the conflict, while the HT approaches it from the perspective of total war and the threat to its aspirations or even its survival. The HT is willing to commit all means necessary, for as long as necessary, to achieve its strategic goals. Compared to the extraregional enemy, the HT stands more willing to absorb higher combatant and noncombatant casualties in order to achieve victory. It will try to influence public opinion in the enemy's homeland to the effect that the goal of intervention is not worth the cost.

4-34. Battlefield victory does not always go to the best-trained, best-equipped, and most technologically advanced force. The collective will of a nation-state or non-state organization encompasses a unification of values, morals, and effort among its leadership, its forces, and its individual members. Through this unification, all parties are willing to individually sacrifice for the achievement of the unified goal. The interaction of military actions and other instruments of power, conditioned by collective will, serves to further define and limit the achievable objectives of a conflict for all parties involved. These factors can also determine the duration of a conflict and conditions for its termination.

NEUTRALIZE TECHNOLOGICAL OVERMATCH

4-35. Against an extraregional force, the HT's forces will forego massed formations, patterned echelonment, and linear operations that would present easy targets for such an enemy. The HT will hide and disperse its forces in areas of sanctuary that limit the enemy's ability to apply his full range of technological capabilities. However, the HT can rapidly mass forces and fires from those dispersed locations for decisive combat at the time and place of its own choosing.

4-36. The HT will attempt to use the physical environment and natural conditions to neutralize or offset the technological advantages of a modern extraregional force. It trains its forces to operate in adverse weather, limited visibility, rugged terrain, and urban environments that shield them from the effects of the enemy's high-technology weapons and deny the enemy the full benefits of his advanced C2 and reconnaissance, intelligence, surveillance, and target acquisition (RISTA) systems.

4-37. The HT can also use the enemy's robust array of RISTA systems against him. His large numbers of sensors can overwhelm his units' ability to receive, process, and analyze raw intelligence data and to provide timely and accurate intelligence analysis. The HT can add to this saturation problem by using deception to flood enemy sensors with masses of conflicting information. Conflicting data from different sensors at different levels (such as satellite imagery conflicting with data from unmanned aerial vehicles) can confuse the enemy and degrade his situational awareness.

4-38. The destruction of high-visibility or unique systems employed by enemy forces offers exponential value in terms of increasing the relative combat power of the HT's forces. However, these actions are not always linked to military objectives. They also maximize effects in the information and psychological arenas. High-visibility systems that could be identified for destruction could include stealth aircraft, attack helicopters, counterbattery artillery radars, aerial surveillance platforms, or rocket launcher systems. Losses among these premier systems may undermine enemy morale, degrade operational capability, and inhibit employment of these weapon systems.

4-39. If available, precision munitions can degrade or eliminate high-technology weaponry. Camouflage, deception, decoy, or mockup systems can degrade the effects of enemy systems. Also, HT forces can employ low-cost GPS jammers to disrupt enemy precision munitions targeting, sensor-to-shooter links, and navigation. Another way to operate on the margins of enemy technology is to maneuver during periods of reduced exposure, those periods identified by a detailed study of enemy capabilities.

4-40. Modern militaries rely upon information and information systems to plan and conduct operations. For this reason, the HT will conduct extensive information attacks and other offensive INFOWAR actions. It could physically attack enemy systems and critical C2 nodes, or conduct "soft" attacks by utilizing computer viruses or denial-of-service activities. Attacks can target enemy military and civilian decisionmakers and key information nodes such as information network switching centers, transportation centers, and aerial platform ground stations. Conversely, HT information systems and procedures should be designed to deny information to the enemy and protect friendly forces and systems with a well-developed defensive INFOWAR plan.

4-41. The HT may have access to commercial products to support precision targeting and intelligence analysis. This proliferation of advanced technologies permits organizations to achieve a situational awareness of enemy deployments and force dispositions formerly reserved for selected militaries. Intelligence can also be obtained through greater use of human intelligence (HUMINT) assets that gain intelligence through civilians or local workers contracted by the enemy for base operation purposes. Similarly, technologies such as cellular telephones are becoming more reliable and inexpensive. It is becoming harder to discriminate between use of such systems by civilian and military or paramilitary actors. Therefore, they could act as a primary communications system or a redundant measure of communication, and there is little the enemy can do to prevent their use.

CHANGE THE NATURE OF CONFLICT

4-42. The HT will try to change the nature of conflict to exploit the differences between friendly and enemy capabilities. To do this, it can take advantage of the opportunity afforded by phased deployment by an extraregional enemy. Following an initial period of regionally-focused conventional operations, the HT will change its operations to focus on preserving combat power and exploiting enemy ROE. This change of operations will present the fewest targets possible to the rapidly growing combat power of the enemy. It is possible that enemy power-projection forces, optimized for a certain type of maneuver warfare, would be ill suited to continue operations. (An example would be a heavy-based projection force confronted with combat in complex terrain.)

4-43. Against early-entry forces, the HT may still be able to use the design it employed in previous operations against regional opponents. However, as the extraregional force builds up to the point where it threatens to overmatch HT forces, the HT is prepared to disperse its forces and employ them in patternless operations that present a battlefield that is difficult for the enemy to analyze and predict.

4-44. The HT may hide and disperse its forces in areas of sanctuary. The sanctuary may be physical, often located in urban areas or other complex terrain that limits or degrades the capabilities of enemy systems. However, the HT may also use moral sanctuary by placing its forces in areas shielded by civilians or close to sites that are culturally, politically, economically, or ecologically sensitive. The HT's forces will defend in sanctuaries, when necessary. However, elements of those forces will move out of sanctuaries and attack when they can create a window of opportunity or when opportunity is presented by physical or natural conditions that limit or degrade the enemy's systems.

4-45. The strengths and weaknesses of an adversary may require other adjustments. The HT will capitalize on interoperability issues among the enemy forces and their allies by conducting rapid actions before the enemy can respond with overwhelming force. If the HT operates near the border of a country with a sympathetic population, it can use border areas to provide refuge or a base of attack for insurgent or other irregular forces. Also, the HT can use terror tactics against enemy civilians or soldiers not directly connected to the intervention as a device to change the fundamental nature of the conflict.

4-46. The HT may have different criteria for victory than the extraregional force—a stalemate may be good enough. Similarly, its definition of victory may not require a convincing military performance. For example, it may call for inflicting numerous casualties to the enemy. The HT's perception of victory may equate to its survival. So the nature of the conflict may be perceived differently in the eyes of the HT versus those of the enemy.

ALLOW NO SANCTUARY

4-47. The HT seeks to deny enemy forces safe haven during every phase of a deployment and as long as they are in the region. The resultant drain on manpower and resources to provide adequate force-protection measures can reduce the enemy's strategic, operational, and tactical means to conduct war and erode his national will to sustain conflict.

4-48. Along with dispersion, decoys, and deception, the HT uses urban areas and other complex terrain as sanctuary from the effects of enemy forces. Meanwhile, its intent is to deny enemy forces the use of such terrain. This forces the enemy to operate in areas where the HT's fires and strikes can be more effective.

4-49. Terror tactics are one of the effective means to deny sanctuary to enemy forces. Terrorism has a purpose that goes well beyond the act itself. The goal is to generate fear. For the HT, these acts are part of the concept of total war. HT-sponsored or -affiliated terrorists or independent terrorists can attack the enemy anywhere and everywhere. The HT's special-purpose forces (SPF) can also use terror tactics and are well equipped, armed, and motivated for such missions.

4-50. The HT is prepared to attack enemy forces anywhere on the battlefield, at overseas bases, at home stations, and even in military communities. It will attack his airfields, seaports, transportation infrastructures, and lines of communications (LOCs). These attacks feature coordinated operations by all available forces, using not just terror tactics, but possibly long-range missiles and weapons of mass destruction (WMD). Targets include not only enemy military forces, but also contractors and private firms involved in transporting troops and materiel into the region. The goal is to present the enemy with a nonlinear, simultaneous battlefield. Striking such targets will not only deny the enemy sanctuary, but also weaken his national will, particularly if the HT can strike targets in the enemy's homeland.

EMPLOY OPERATIONAL EXCLUSION

4-51. The HT will apply operational exclusion to selectively deny an extraregional force the use of or access to operating bases within the region or near it. In doing so, it seeks to delay or preclude military operations by the extraregional force. For example, through diplomacy, economic, or political connections, information campaigns, and/or hostile actions, the HT might seek to deny the enemy the use of bases in other foreign nations. It might also attack population and economic centers for the intimidation effect, using long-range missiles, WMD, or SPF.

4-52. Forces originating in the enemy's homeland must negotiate long and difficult air or surface LOCs merely to reach the region. Therefore, the HT will use any means at its disposal to also strike the enemy forces along routes to the region, at transfer points en route, at aerial and sea ports of embarkation (APOEs and SPOEs), and even at their home stations. These are fragile and convenient targets in support of transition and adaptive operations.

EMPLOY OPERATIONAL SHIELDING

4-53. The HT will use any means necessary to protect key elements of its combat power from destruction by an extraregional force, particularly by air and missile forces. This protection may come from use of any or all of the following:
- Complex terrain.
- Noncombatants.
- Risk of unacceptable collateral damage.
- Countermeasure systems.
- Dispersion.
- Fortifications.
- INFOWAR.

4-54. Operational shielding generally cannot protect the entire force for an extended time period. Rather, the HT will seek to protect selected elements of its forces for enough time to gain the freedom of action necessary to pursue its strategic goals.

Chapter 5
Hybrid Threat Tactics

The Hybrid Threat (HT) possesses a wide range of options for executing tactical actions. This chapter explores the concepts behind those options.

TACTICAL CONCEPTS

5-1. Initiative and mobility characterize tactics the HT would use while establishing and preserving bases in which to train, self-sustain, prepare for future missions, and evolve organizational capability. Concurrently, collective tactical actions can have strategic consequences of denying an enemy a secure area or making it politically untenable to remain. Actions are aimed at keeping an enemy physically and psychologically stressed from constant harassment and disruption when a distinct defeat or destruction of an enemy is not practical.

5-2. Tactical actions can encompass a range of activities that can include the following:
- Collection of intelligence.
- Coercion for fiscal or logistic support.
- Assassination of designated enemy leaders or officials.
- Sabotage by small loosely affiliated groups of irregular forces.
- More traditional major offensive and defensive actions between regular military forces.

SYNERGY OF REGULAR AND IRREGULAR FORCES

5-3. The HT understands that the environment that would produce the most challenges to U.S. forces is one in which conventional military operations occur in concert with irregular warfare. The HT's concept is not just one of making do with what is available, but is primarily one of deliberately created complexity.

5-4. Each component of the HT brings a capability to bear. The synergy of these capabilities is not to be understated. Operational environments (OEs) by their very nature provide a myriad of complexities across all the operational variables. The HT seeks to introduce additional complexity through the use of an ever-shifting array of forces, technologies, and techniques.

INFORMATION WARFARE AS A KEY WEAPON SYSTEM

5-5. HT tactical actions will be often be designed to achieve information warfare (INFOWAR) objectives rather than purely military ones. Information and its management, dissemination, and control have always been critical to the successful conduct of tactical missions. Given today's tremendous advancements in information and information systems technology, this importance is growing in scope, impact, and sophistication. The HT recognizes the unique opportunities that INFOWAR gives tactical commanders. Therefore, it continuously strives to incorporate INFOWAR activities in all tactical missions and battles.

5-6. INFOWAR may help degrade or deny effective enemy communications and blur or manipulate the battlefield picture. In addition, INFOWAR helps the HT achieve the goal of dictating the tempo of combat. Using a combination of perception management activities, deception techniques, and electronic warfare (EW), the HT can effectively slow or control the pace of battle. For example, the HT may selectively destroy lucrative enemy targets. It could also orchestrate and execute a perception management activity that weakens the enemy's international and domestic support, causing hesitation or actual failure of the operation. It executes deception plans to confuse the enemy and conceal intentions. More traditional EW activi-

Chapter 5

ties also contribute to the successful application of INFOWAR at the tactical level by challenging and/or weakening the enemy's quest for information dominance.

5-7. INFOWAR also supports the critical mission of counterreconnaissance at the tactical level. The HT constantly seeks ways to attack, degrade, or manipulate the enemy's reconnaissance, intelligence, surveillance, and target acquisition (RISTA) capabilities. All enemy target acquisition systems and sensors are potential targets.

COMPLEX BATTLE POSITIONS

5-8. The HT reduces exposure to enemy standoff fires and RISTA by utilizing complex battle positions (CBPs) and cultural standoff. CBPs are designed to protect the units within them from detection and attack while denying their seizure and occupation by the enemy. Commanders occupying CBPs intend to preserve their combat power until conditions permit offensive action. In the case of an attack, CBP defenders will engage only as long as they perceive an ability to defeat aggressors. Should the defending commander feel that his forces are decisively overmatched, he will attempt a withdrawal in order to preserve combat power.

5-9. CBPs have the following characteristics that distinguish them from simple battle positions (SBPs):
- Limited avenues of approach. (CBPs are not necessarily tied to an avenue of approach.)
- Avenues of approach are easily observable by the defender.
- 360-degree fire coverage and protection from attack. (This may be due to the nature of surrounding terrain or engineer activity such as tunneling.)
- Engineer effort prioritizing camouflage, concealment, cover, and deception (C3D) measures; limited countermobility effort which might reveal the CBP location.
- Large logistics caches.
- Sanctuary from which to launch local attacks.

5-10. C3D measures are critical to the success of a CBP, since the defender generally wants to avoid enemy contact. Additionally, forces within a CBP will remain dispersed to negate the effects of precision ordinance strikes. Generally, once the defense is established, non-combat vehicles will be moved away from troop concentrations to reduce their signature on the battlefield.

5-11. Cultural standoff is the fact that protection from enemy weapon systems can be gained through actions that make use of cultural differences to prevent or degrade engagement. Examples of cultural standoff are—
- Using a religious or medical facility as a base of fire.
- Firing from within a crowd of noncombatants.
- Tying prisoners in front of battle positions and onto combat vehicles.

SYSTEMS WARFARE

5-12. The HT will disaggregate enemy combat power by destroying or neutralizing vulnerable single points of failure in enemy warfighting functions. A system is a set of different elements so connected or related as to perform a unique function not performable by the elements or components alone. The essential ingredients of a system include the components, the synergy among components and other systems, and some type of functional boundary separating it from other systems. Therefore, a "system of systems" is a set of different systems so connected or related as to produce results unachievable by the individual systems alone. The HT views the OE, the battlefield, its own instruments of power, and an opponent's instruments of power as a collection of complex, dynamic, and integrated systems composed of subsystems and components.

5-13. Systems warfare serves as a conceptual and analytical tool to assist in the planning, preparation, and execution of warfare. With the systems approach, the intent is to identify critical system components and attack them in a way that will degrade or destroy the use or importance of the overall system.

5-14. The primary principle of systems warfare is the identification and isolation of the critical subsystems or components that give the opponent the capability and cohesion to achieve his aims. The focus is on the disaggregation of the system by rendering its subsystems and components ineffective. While the aggregation of these subsystems or components is what makes the overall system work, the interdependence of these subsystems is also a potential vulnerability.

ADAPTING BY FUNCTION

5-15. The HT will choose the most effective option for executing each combat function, without regard to original purpose, laws of war, or military hierarchy. For example, a child on a street corner with a cell phone may be the most effective means of providing early warning to the leaders involved in a tactical action. If so, the HT will employ that option, even if more sophisticated or expensive RISTA devices or techniques are available.

5-16. The HT will typically acquire a capability to permit it to act with freedom with respect to its natural, regional enemies. These capabilities will be adapted to exploit their pertinent characteristics to best advantage against enemy forces.

FUNCTIONAL TACTICS

5-17. The HT employs functional tactics. It determines the functions that need to be performed as part of an action to bring about its success. Then it allocates appropriate actors to each function and synchronizes the effort.

5-18. A number of different functions must be executed each time an HT force attempts to accomplish a mission. An HT commander identifies the specific functions he intends his various subordinate forces or elements to perform. The functions do not change, regardless of where the force or element might happen to be located on the battlefield. However, the function of a particular force or element may change during the course of the battle. While the various functions required to accomplish any given mission can be quite diverse, they can be broken down into two very broad categories: action and enabling.

Note. In larger groupings of forces, HT commanders refer to the subordinates performing various functions as *forces*. In smaller groupings, commanders call them *elements*.

ACTION FUNCTIONS

5-19. The *action* function is performed by the set of capabilities that actually accomplish a given mission. One part of the unit or grouping of units conducting a particular *action* is normally responsible for performing the primary function or task that accomplishes the goal or objective of that *action*. In most general terms, therefore, that part can be called the *action force* or *action element*. In most cases, however, the higher commander will give the action force or element a more specific designation that identifies the specific function it is intended to perform, which equates to achieving the objective of the higher command's mission.

Examples

5-20. For example, if the objective of the action is to conduct an *assault*, the element designated to complete that action is the *assault element*. In larger offensive actions, an action force that completes the primary offensive mission by *exploiting* a window of opportunity created by another force is called the *exploitation force*. In defensive actions, the unit or grouping of units that performs the *main defensive* mission in the battle zone is called the *main defense force* or *main defense element*. However, in a maneuver defense, the main defensive action is executed by a combination of two functional forces: the *contact force* and the *shielding force*.

5-21. If the HT objective is to destroy a city with a weapon of mass destruction (WMD), then the WMD is performing the action function. In another instance, the HT objective could be to seize a friendly capital city and it might employ a WMD in another area to force a response by enemy forces that leaves the capital

Chapter 5

exposed. In that case, the force used to seize the capital is performing the action function, and the WMD is performing a different (enabling) function.

Note. In defensive actions, there may be a particular unit or grouping of units that the HT commander wants to be *protected* from enemy observation or fire, to ensure that it will be available after the current battle or operation is over. This is designated as the *protected force*. This protected force could become the action force in a subsequent battle or operation.

Extended Example

5-22. An extended example of an HT raid follows to illustrate the action function. (See figure 5-1.) In this example, an insurgent force conducts a raid on an enemy combat outpost. In this raid, the assault element executes the action function, exploiting the breach created by the breach element (an enabling function).

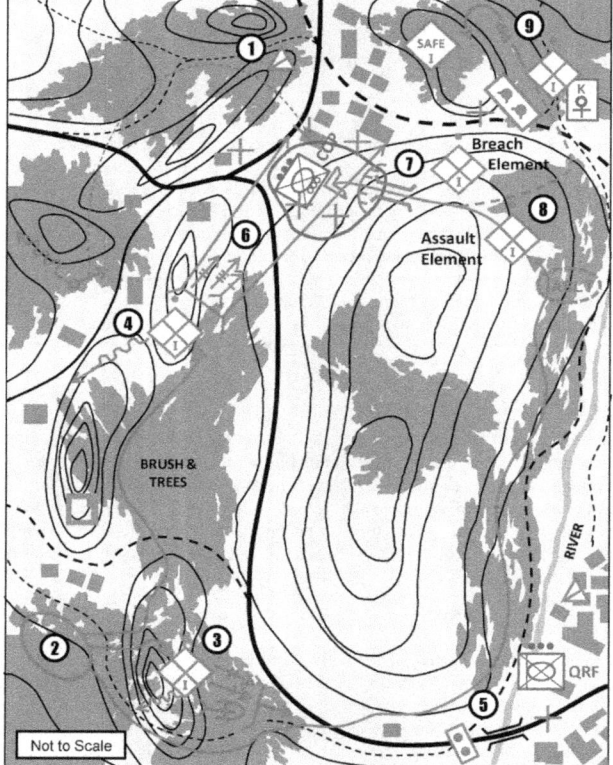

Figure 5-1. Insurgent raid on a combat outpost (example)

5-23. Local insurgents conduct reconnaissance and surveillance ① on enemy combat outposts in battle zone. Observers record and report actions such as scheduled deliveries, work shift changes, identification procedures and other routines. Preplanned feints or demonstrations have allowed collection on the complexity of the security around the target that includes—
- Manning strength of the combat outposts and its weapon systems.
- Reaction time of response units.
- Any hardening of structures, barriers, or sensors.
- Personnel, package, and vehicle screening procedures.
- Type and frequency of emergency reaction drills.

This intelligence identifies an attack target that will most likely lead to insurgent success by avoiding substantial enemy security measures. Observation continues before, during, and after the raid by insurgent elements not involved in the assault or direct support.

5-24. Once a specific combat outpost has been selected as the target, insurgent forces plan and rehearse actions in an assembly area ②. In a coordinated sequence of actions, insurgent elements move to designated positions that in this example are—
- An attack by fire position.
- A support by fire position.
- An assault position.

5-25. The attack by fire element ③ occupies hide positions in the vicinity of fighting positions. The support by fire element ④ moves along a separate route to a position overwatching the combat outpost and occupies a nearby hide position. Light mortars are emplaced away from the direct fire systems of medium and heavy machineguns and antitank weapons. Indirect and direct fires intend to fix the combat outpost and disrupt the enemy defense.

5-26. The attack by fire element emplaces an antitank (AT)-capable series of mines near the bridge ⑤ to delay any quick reaction force (QRF) from impacting on the raid at the combat outpost. Insurgents rally at the attack by fire position ③ and wait. This choke point is also planned for indirect fire support.

5-27. The assault and breach elements move via a covered and concealed route, continue through an assault position without halting, and position for the attack. The action of the assault element is enabled by a breach element, which breaches the identified weak point in the combat outpost defensive position and secures the breach and immediate flanks for the assault element to attack into the combat outpost.

5-28. On order, the support by fire element initiates the raid ⑥ with direct and indirect fires on the combat outpost. As the support by fire element shifts fires and continue to fix the combat outpost, the breach element ⑦ moves forward and breaches the defensive perimeter. The assault element supports the breach point with fires and attacks through the breach ⑧ once the penetration is confirmed by the breach element.

5-29. The attack by fire element ③ detonates AT mines at the bridge exit to delay the QRF. Medium and light machinegun fire and antitank guided missile fire block the QRF ⑤ while the assault is in progress.

5-30. The assault element attacks into the combat outpost to defeat the enemy force and seize hostages. The assault and breach elements withdraw quickly with hostages, under pressure from the combat outpost perimeter. The withdrawal route through restricted terrain is prepared with antipersonnel (AP) mines and is planned for indirect fires of high explosive and smoke to delay pursuit by enemy forces. The insurgents in this group break contact ⑨ and move to a safe haven with their hostages.

5-31. The attack by fire element ③ disengages from the firefight at the bridge and exfiltrates to a safe haven. The support by fire element disengages in a phased withdrawal of direct fire systems. The mortar section continues to provide harassment indirect fires and disengages when the direct fires elements have departed their positions and exfiltrate to a safe haven. Observation throughout the attack may be videotape recorded, along with written observer reports, to produce after-action analysis and lessons learned for effective future combat actions.

Chapter 5

ENABLING FUNCTIONS

5-32. The *enabling* function is performed by a set of capabilities that acts to assist those capabilities performing the action function. In relation to the force(s) or element(s) conducting the action function, all other parts of the organization or grouping of organizations conducting an action provide *enabling* functions of various kinds. In most general terms, therefore, each of these parts can be called an *enabling force* or *enabling element*. However, each subordinate force or element with an enabling function can be more clearly identified by the specific function it performs.

Examples

5-33. For example, a force that enables by *fixing* enemy forces so they cannot interfere with the primary action is a *fixing force*. Likewise, an element that creates a *breach* that enables an assault element to assault the enemy forces on the far side of an obstacle is a *breach element*.

5-34. In larger offensive actions, one force can enable another by conducting an *assault* that enables another force to exploit the effects of that assault in order to accomplish the primary objective. Thus, that type of enabling force can be called the *assault force*. In this case, the force that conducts the initial assault is not the one that is actually intended to achieve the objective of the higher command's mission. The role of the assault force is to create an opportunity for another force—the exploitation force—to accomplish the objective. Thus, the assault force, conducting the first part of a two-part offensive action, acts as an enabling force. In order to create a window of opportunity for the exploitation force to succeed, the assault force may be required to operate at a high degree of risk and may sustain substantial casualties. However, other types of enabling forces or elements may not even need to make contact with the enemy.

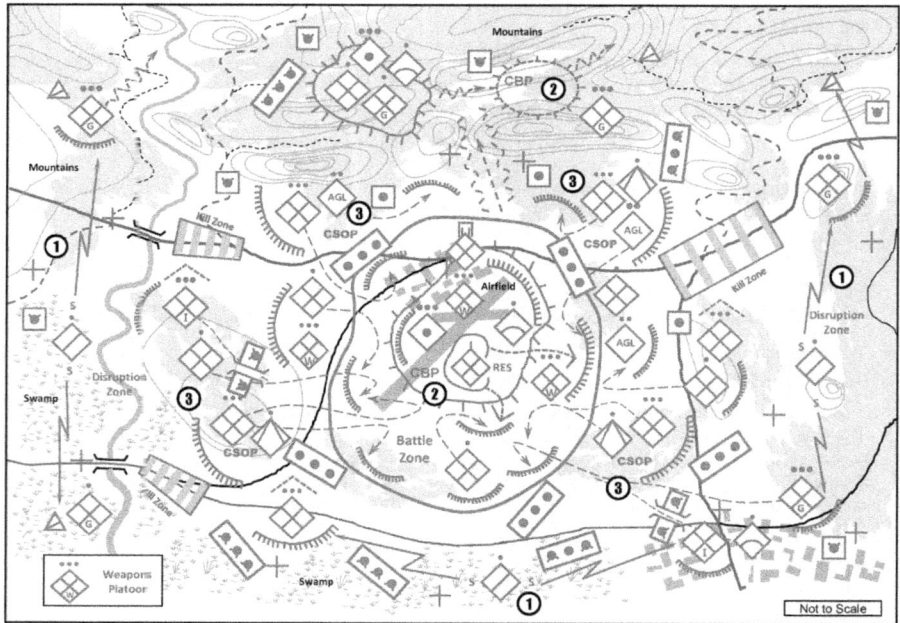

Figure 5-2. HT area defense (example)

Hybrid Threat Tactics

5-35. If the mission is to enter an enemy base and set off an explosive device, an enabling function would be to penetrate the perimeter defenses of the base or to assist in the infiltration of the element emplacing the device. In the defense, an enabling function might be to counterattack to restore a portion of the area of responsibility (AOR) to HT control.

> *Note.* A unit or group of units designated as a particular functional force or element may also be called upon to perform other, more specific enabling functions. Therefore, the function of that force or element, or part(s) of it, may be more accurately described by a more specific functional designation. For example, a *disruption force* generally *disrupts*, but also may need to *fix* a part of the enemy forces. In that case, the entire disruption force could become the *fixing force*, or parts of that force could become *fixing elements*.

Extended Example

5-36. An example of an HT area defense follows to illustrate the enabling function. (See figure 5-2.) In this example, a hybrid force of guerrillas, insurgents, and military units conducts an area defense to protect their leadership from attack or capture.

5-37. Guerrilla forces, reconnaissance teams, and insurgents ① work together as a disruption force to force the enemy off his timetable and prevent his gaining accurate information about HT dispositions. This provides the defenders with the ability to reposition to best respond to enemy attack. CBPs ② protect the HT leadership by preventing the enemy from discovering and attacking them. Combat security outposts (CSOPs) ③ perform the enabling function of providing time for exfiltration should the enemy prove able to defeat forces in the initial CBP.

Enabling by Disruption, Security, and Fixing

5-38. Three specific types of enabling function are so common as to warrant additional attention. These are disruption, security, and fixing. The following sections describe these functions and provide examples.

Disruption

5-39. *Disruption* forces or elements operate in what the HT refers to as the disruption zone. They can—

- Disrupt enemy preparations or actions.
- Destroy or deceive enemy reconnaissance.
- Begin reducing the effectiveness of key components of the enemy's combat system.

This function is so common that it appears in many examples of other functions in this chapter, as well as in the scenario blueprint examples in appendix A. Therefore, no specific examples are included here.

Security

5-40. The *security* function is performed by a set of capabilities that acts to protect other capabilities from observation, destruction, or becoming fixed. Security is provided by isolating the battlefield from enemy elements that could alter the outcome. This can be accomplished by providing early warning and reaction time or actively delaying or destroying arriving enemy forces.

5-41. An extended example follows to illustrate the function of security. (See figure 5-3 on page 5-8.) Insurgent forces in company strength have agreed to provide security for a local criminal organization at a drug manufacturing site ① in a remote area contested by two sovereign states. Insurgent reconnaissance identifies that the drug site is not located along a logical enemy avenue of approach and is a satisfactory location for development of a CBP defense. The CBP ② is configured to protect elements and materiel within the defensive perimeter and support zone from detection and attack, and deny their seizure and occupation by the enemy. Although the HT may be aware of possible and likely coalition approaches into the guerrilla battle zone, the primary purpose of this CBP is to provide sanctuary. If necessary, this protection for the temporary drug manufacturing and storage shelters activates a defense or delay.

Chapter 5

5-42. Security observation posts (OPs) monitor possible enemy access routes ③ into the CBP. The insurgents identify most likely and most dangerous directions of attack and prioritize the CBP defense to them. Indirect and direct fires are planned for kill zones that support the defense in key locations such as the only ford across the river. Confirmed as a difficult ford crossing, the near bank ④ has AP mines emplaced to slow any attack. The disruption force ③ uses AP mines, trip wires, and abatis to provide early warning and impede any enemy maneuver toward the CBP. On order, indirect fires into a kill zone or preplanned targets supplement this early disruption of any attack. The other possible direction of attack is an exposed deep swamp ⑤ with excellent fields of direct fire, and the insurgents plan a kill zone there with indirect fires.

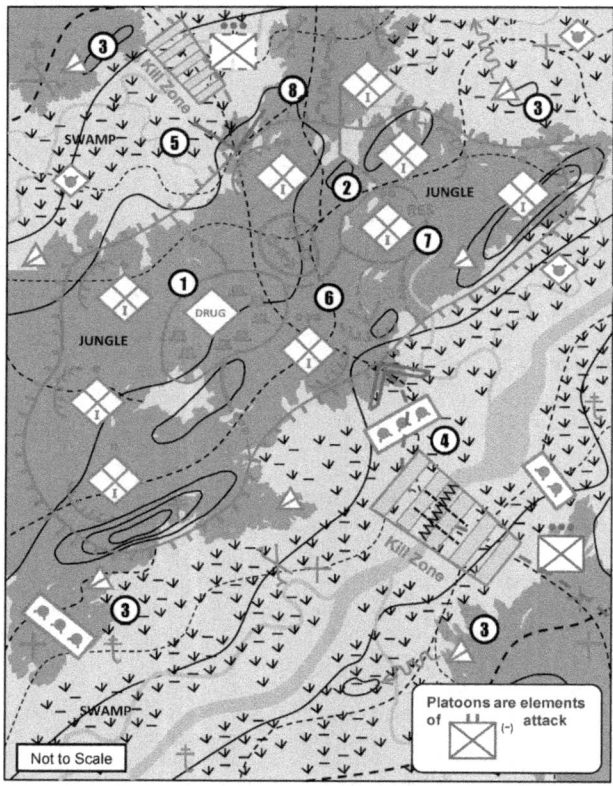

Figure 5-3. Insurgents provide security for a drug manufacturing site (example)

5-43. Enemy forces have elements of a light infantry battalion operating in conjunction with drug enforcement officials. The jungle terrain, marshland, and the dense vegetation make movement off the limited trail systems almost inaccessible. Fighting positions orient on the most likely and most dangerous direction of attack into the CBP defense ⑥.

5-44. The insurgent reserve ⑦ is positioned near the CBP support zone for rapid response to preselected battle positions. Light and medium mortars are positioned to provide immediate indirect fires against any direction of attack.

5-45. The insurgent force is also prepared for the possibility that an enemy attack is successful in penetrating into the battle zone and the drug manufacturing site is in jeopardy of capture. In that case, the insurgent force conducts a delay to allow evacuation of drug material along a preselected exfiltration route ⑧. It has also emplaced nuisance AP mines along that route to delay any pursuing enemy forces. Automatic rifle teams using a series of hasty defense bounding overwatch positions can impose further delay.

5-46. Security OPs will remain in position in the disruption zone when bypassed unknowingly by enemy forces. These observers will assist in calls for indirect fire to disrupt enemy forces in the assault on the CBP. Insurgent forces in the CBP conduct an orderly delay and disengage, on order, to exfiltrate in small groups to rendezvous points in the area. Observations before and during an attack, and reports from any stay-behind insurgent element will be collated in after-action analysis and lessons learned for effective future combat actions.

Fixing

5-47. The *fixing* function is performed by a set of capabilities that acts to prevent opposing capabilities from interfering with mission accomplishment. Fixing is accomplished when a part of the enemy force does not participate in actions that could lead to the failure of the HT course of action. This can be accomplished in a variety of ways, including—
- Suppressing a force with fires.
- Deceiving it with INFOWAR.
- Forcing it to conduct consequence management.
- Involving it in a firefight away from the main action.
- Restricting its movement with countermobility effects.
- Depriving it of logistics resources.

5-48. An extended example of an HT raid follows to illustrate the function of fixing. (See figure 5-4 on page 5-10.) In this example, a guerrilla force conducts a raid to secure a high-value target.

Chapter 5

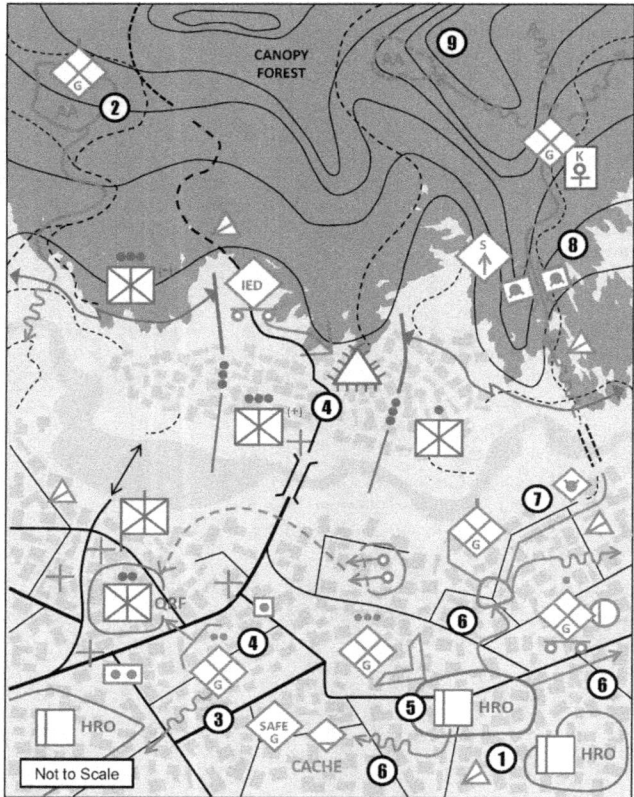

Figure 5-4. Guerrilla force conducts a raid (example)

5-49. Local guerrillas maintain regular surveillance ① of the operations of a humanitarian relief organization (HRO) in an urban area near the guerrilla sanctuary. Observers record and report actions along major routes in and out of the target area, and focus attention on a particular HRO support site ⑤ in a warehouse section of the town. Intelligence priorities of effort include—

- Scheduled deliveries.
- Work shift changes.
- Contract security at the site.
- Identification procedures.
- Other routines at this HRO site and urban activities in the vicinity.

5-50. Guerrillas know that this urban complex is defended by a motorized infantry company with no other enemy ground maneuver forces nearby. Most of the infantry company combat power is oriented north of the river line to protect the only multi-span bridge and ferry site in the area. Enemy defense concentrates around a combat outpost on the one trafficable north-south road. Guerrilla surveillance confirms that other company elements north of the river are conducting screening patrols on an intermittent basis.

5-51. The HRO operation is actually several organizations with loosely affiliated field offices located throughout the urban center. None of the HRO activities want military forces near their support sites due to guerrilla propaganda that claims the HRO missions are part of a military coalition and governmental ploy to control the local populace with food, water purification, health services, and general supplies.

5-52. A small enemy QRF is positioned south of the river for rapid response to either the bridge and ferry site or the general urban area. Preplanned guerrilla feints or demonstrations have collected intelligence on the security around the target that includes—
- Manning strength, transportation, and weapon systems of the QRF.
- Reaction time and priority routes of QRF response.
- Any hardening of combat outpost barriers.
- Any changes in personnel, package, and vehicle screening procedures during peak periods of traffic flow.

5-53. Reconnaissance has identified an unmarked ford site for infiltration by guerrilla elements into the urban complex. After rehearsing actions ② in an assembly area, guerrilla elements position in hide positions ③ and a safe haven near the HRO site and QRF. A vehicle-borne improvised explosive device initiates the raid when it rams the combat outpost checkpoint ④ and detonates. Simultaneously, the attack by fire element fixes ④ the QRF with medium machine gun and light mortar fires. Several AT mines are positioned in a hasty group to impede the probable motorized QRF approaches to the HRO target site.

5-54. The guerrilla main effort assaults the HRO site ⑤, seizes control of the warehouse and HRO personnel, and moves to designated supply pallets. Guerrillas quickly load supplies into an HRO truck and drive to a rendezvous on the outskirts of the urban area to transfer supplies ⑥ and disperse into the populace. Other dismounted guerrilla elements move designated supplies to several caches within the urban complex. An additional group with HRO hostages rallies with other guerrillas and exfiltrates north of the river. The attack by fire element disengages, on order, from the QRF and exfiltrates to a safe haven. Surveillance OPs assist exfiltration and control passage lanes through nuisance AP mines ⑦ and coordinate a sniper to delay ⑧ any pursuing forces in the group of AP mines.

5-55. Guerrilla elements divide into small groups and exfiltrate to a tentative assembly area ⑨ or designated safe havens. Observation continues before, during, and after the raid by guerrilla elements not involved in the combat assault or direct support. This continuous reconnaissance and surveillance complements after-action analysis and lessons learned for effective future combat actions.

Other Enabling Functions

5-56. The commander may designate a subordinate unit or grouping to conduct a *deception* action (such as a demonstration or feint). This unit or grouping is, therefore, a *deception force* or *deception element*. Its function is to lead the enemy to act in ways prejudicial to enemy interests or favoring the success of an HT action force or element.

5-57. A commander may also designate some subordinates to perform various *support* functions. These *support elements* can provide the following types of support:
- Perform support by fire (in which case it can be called more specifically a *support by fire element*).
- Provide combat or combat service support.
- Provide command and control functions.

5-58. At a commander's discretion, some forces or elements may be held out of initial action, in *reserve*, pending determination of their specific function, so that he may influence unforeseen events or take advantage of developing opportunities. These are designated as *reserves (reserve force* or *reserve element)*. If and when such units are subsequently assigned a mission to perform a specific function, they receive the appropriate functional force or element designation. For example, a reserve force in a defensive operation might become the *counterattack* force.

This page intentionally left blank.

Chapter 6
Hybrid Threat Organizations

The Hybrid Threat (HT) tailors its organizations to the required missions and functions. It determines the functions that must be performed in order to successfully accomplish its goals. Then it builds teams and organizations to execute those functions without regard to traditional military hierarchy, the law of war, or rules of engagement.

TASK-ORGANIZING

6-1. The HT will task-organize forces in a fashion that matches its available resources to its goals. Task organizations will often include more than purely military formations. The HT's regular military and irregular components are tailored forces depending on training requirements. FM 7-100.4 provides a baseline of organizational size, equipment, and weapons. Its organizational directories provide a very detailed listing of personnel and equipment. For some training requirements, the opposing force (OPFOR) order of battle (OB) might not need to include personnel numbers. Trainers and exercise planners can extract the appropriate pages from the organizational directories and tailor them by eliminating the detail they do not need and adding the necessary units from other pages to develop the required task organization. For more detail on organizations, see FM 7-100.4, which introduces baseline organizational structures of a flexible, thinking, and adaptive OPFOR.

6-2. The baseline organizations presented in the organizational directories of FM 7-100.4 are intended to be tailored and task-organized in a manner that is appropriate for the training objectives. Depending on the training requirement, the OPFOR may be a large, medium, or small force. Its technology may be state-of-the-art, relatively modern, obsolescent, obsolete, or an uneven combination of these categories. Its ability to sustain operations may be limited or robust.

MILITARY ORGANIZATIONS

6-3. Regular military organizations of the HT will present conventional and unconventional capabilities. This TC is part of the TC 7-100 series, which includes OPFOR doctrine, organization, and equipment for trainers and educators to tailor specified threats for U.S. Army training requirements.

6-4. In the regular military forces of a nation-state that is part of the HT, six services generally comprise the armed forces. These include the Army, Navy, Air Force (which includes the national-level Air Defense Forces), Strategic Forces (with long-range rockets and missiles), Special-Purpose Forces (SPF) Command, and Internal Security Forces. The Internal Security Forces may be subordinate to the Ministry of the Interior rather than to the Ministry of Defense. The armed forces field some reserve component forces in all services, but most reserve forces are Army forces. In time of war, command and control relationships among state ministries may be consolidated for regular, reserve, militia, and other paramilitary-type armed forces, all under the Supreme High Command (SHC).

6-5. Baseline OPFOR organizations described in FM 7-100.4 do not constitute an OPFOR OB. Rather, they provide a framework from which trainers can develop a specific OPFOR OB appropriate for their particular training requirements. Within this framework, training scenario writers and exercise designers have considerable flexibility in determining what the OPFOR actually has in capabilities or limitations at a given point in time or a given location. In some cases, an organization taken straight from the OPFOR administrative force structure (AFS) in FM 7-100.4 may meet the requirements for a particular U.S. Army training environment. In most cases, however, task-organizing an OPFOR organization is appropriate in order to

Chapter 6

portray the correct array of OPFOR units and equipment for stressing the mission essential task list (METL) of U.S. units in a particular training environment.

SPECIAL-PURPOSE FORCES COMMAND

6-6. As part of an OPFOR, the SPF Command includes both SPF units and elite commando units. Four of the five other service components of the armed forces also have their own SPF. There are Army, Navy, and Air Force SPF. The Internal Security Forces also have their own SPF units. These service SPF normally remain under the control of their respective services or a joint operational or theater command. However, SPF from any of these service components could become part of joint SPF operations in support of national-level requirements. The SPF Command has the means to control joint SPF operations as required.

6-7. Any SPF units from the SPF Command or from other service components' SPF that have reconnaissance or direct action missions supporting strategic-level objectives or intelligence requirements would normally be under the direct control of the SHC or under the control of the SPF Command, which reports directly to the SHC. Also, any service SPF units assigned to joint SPF operations would temporarily come under the control of the SPF Command or perhaps the SHC. Most of the service SPF units are intended for use at the operational level. Thus, they can be subordinate to operational-level commands even in the AFS. In peacetime and in garrisons, SPF of both the SPF Command and other services are organized administratively into SPF companies, battalions and brigades.

6-8. In time of war, some SPF units from the SPF Command or from the Army, Navy, Air Force, or Internal Security Forces SPF may remain under the command and control of their respective service headquarters. However, some SPF units also might be allocated to operational or even tactical level commands during the task-organizing process. (See FM 7-100.4 for additional discussion on the strategic to tactical levels of SPF.)

6-9. Regardless of the parent organization in the AFS, SPF normally infiltrate and operate as small teams. When deployed, these teams may operate individually, or they may be task-organized into detachments. The terms *team* and *detachment* indicate the temporary nature of the groupings. In the course of an operation, teams can leave a detachment and join it again. Each team may in turn break up into smaller teams (of as few as two men) or, conversely, come together with other teams to form a larger team, depending on the mission. At a designated time, teams can join up and form a detachment (for example, to conduct a raid), which can at any moment split up again. This whole process can be planned before the operation begins, or it can evolve during the course of an operation.

INTERNAL SECURITY FORCES

6-10. Internal security forces are part of an OPFOR structure for operations against internal threats to the state. In peacetime, the Chief of Internal Security heads the forces within the Ministry of the Interior that fall under the general label of "internal security forces." Most of the internal security forces are uniformed and use military ranks and insignia similar to those of the other services of the nation-state armed forces. Among the internal security forces, border guard, security, and SPF units most closely resemble regular military units of other services of the armed forces. However, units from the General Police directorate and Civil Defense Directorate can also perform military-like roles.

6-11. During wartime, some or all of the internal security forces from the Ministry of the Interior may become a sixth service component of the Armed Forces, with the formal name "Internal Security Forces." Internal Security Forces can be allocated to a theater command or to a task-organized operational or tactical level military command that is capable of controlling joint or interagency operations. In such command relationships or when missions share a common area of responsibility (AOR) with a military organization, units of the Internal Security Forces send liaison teams to represent them in the military organization's staff.

6-12. Various types on non-state actors might be part of the HT, affiliated with it, or support it in some manner. Even those internal security forces that do not belong to the HT, or support it directly or willingly, could be exploited or manipulated by the HT to support its objectives.

Hybrid Threat Organizations

General Police Directorate

6-13. The General Police Directorate has responsibility for national, district, and local police. In some circumstances, police forces at all three levels operate as paramilitary forces. They can use military-type tactics, weapons, and equipment. National Police forces include paramilitary tactical units that are equipped for combat, if necessary. These uniformed forces may represent the equivalent of an infantry organization in the regular armed forces.

6-14. Within the various national- and district-level police organizations, the special police are the forces that most resemble regular armed forces in their organization, equipment, training, and missions. Because some special police units are equipped with heavy weapons and armored vehicles, they can provide combat potential to conduct defensive operations if required. Special police units could be expected to supplement the armed forces.

Civil Defense Directorate

6-15. The Civil Defense Directorate comprises a variety of paramilitary and nonmilitary units. While the majority of Civil Defense personnel are civilians, members of paramilitary units and some staff elements at the national and district levels hold military ranks. Civil Defense paramilitary units are responsible for the protection and defense of the area or installation where they are located. Even the nonmilitary, civil engineering units can supplement the combat engineers of the armed forces by conducting engineer reconnaissance, conducting explosive ordnance disposal, and providing force-protection construction support and logistics enhancements required to sustain military operations.

RESERVES AND MILITIA

6-16. Although all six services can field some reserve forces, most of the reserve forces are Army forces. All militia forces belong to the Army component. Overall planning for mobilization of reserves and militia is the responsibility of the state and its Organization and Mobilization Directorate of the General Staff. Each service component headquarters would have a similar directorate responsible for mobilization of forces within that service. Major geographical commands (and other administrative commands at the operational level and higher) serve as a framework for mobilization of reserve and militia forces.

6-17. During mobilization, some reserve personnel serve as individual replacements for combat losses in active units. Others fill positions, including professional and technical specialists, that were left vacant in peacetime in deference to requirements of the civilian sector. However, reservists also man reserve units that are mobilized as units to replace other units that have become combat-ineffective or to provide additional units necessary for large, sustained operations.

6-18. Like active force units, most mobilized reserve and militia units do not necessarily go to war under the same administrative headquarters that controlled them in peacetime. Rather, they typically become part of a task-organized operational- or tactical-level fighting command tailored for a particular mission. In most cases, the mobilized reserve units would be integrated with regular military units in such a fighting command. In rare cases, however, a reserve command at division level or higher might become a fighting command or serve as the basis for forming a fighting command based partially or entirely on reserve forces.

INSURGENT ORGANIZATIONS

6-19. Insurgent organizations have no regular, fixed table of organization and equipment structure. The mission, environment, geographic factors, and many other variables determine the configuration and composition of each insurgent organization and its subordinate cells. Their composition varies from organization to organization, mission to mission, environment to environment. The structure, personnel, equipment, and weapons mix all depend on specific mission requirements, as does the size, specialty, number, and type of subordinates.

6-20. There are several factors that differentiate the structure and capability of an insurgent organization from the structure and capability of a guerrilla organization. Since the insurgent organization is primarily a

Chapter 6

covert organization, it typically has a cellular, networked structure. By comparison, guerrilla organizations often reflect a military structure such as battalion, company, platoon, or squad.

6-21. Insurgent organizations generally do not have some of the heavier and more sophisticated equipment that guerrilla organizations can possess. Weapons of the insurgents are generally limited to small arms, antitank grenade launchers, and improvised explosive devices (IEDs). There may be some crew-served weapons such as the 82-mm mortar or 107-mm single-tube rocket launcher. In the event the insurgents require heavier weapons or capabilities, they might obtain them from guerrillas, or the guerrilla organization might provide its services depending on the relationship between the two organizations at the time.

6-22. Insurgent organizations are irregular forces. The baseline insurgent organizations in the FM 7-100.4 organizational directories represent the default setting for a typical insurgent organization. If an OPFOR OB has more than one local insurgent organization, no two insurgent organizations should look exactly alike. Trainers and training planners should vary the types and numbers of cells to reflect the irregular nature of such organizations. The FM 7-100.4 baseline array of possible cells for various functions is arranged in a line-and-block chart for convenience. However, they would typically be task-organized in a network-type structure, as shown in the example of a local insurgent organization in figure 6-1.

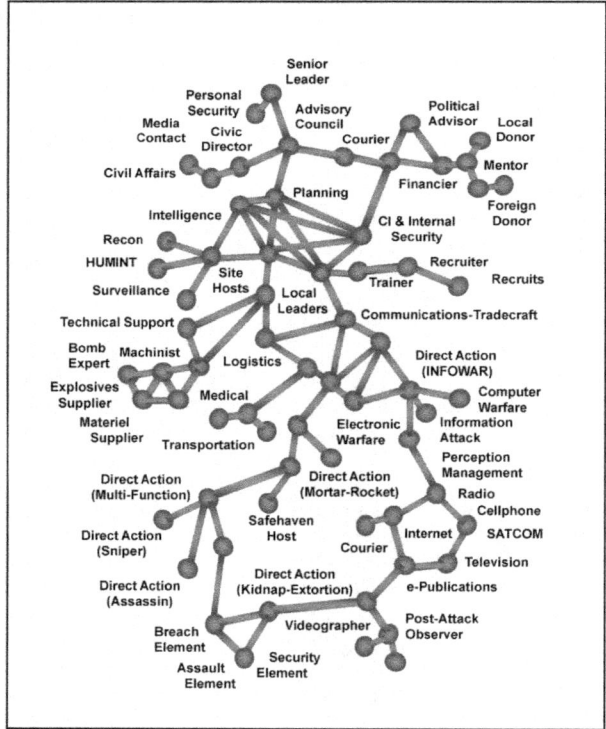

Figure 6-1. Local insurgent organization (example)

6-23. In FM 7-100.4, the baseline organization charts and equipment lists for individual cells include many notes on possible variations in organization or in numbers of people or equipment within a given organization. When developing an OB for a specific insurgent organization for use in training, users may exercise some latitude in the construction of cells. Some cells might need to be larger or smaller than the "default"

setting found in the organizational directories. Some entire cells might not be required, and some functional cells might be combined into a single cell performing multiple functions. However, trainers and training planners would need to take several things into consideration in modifying the "default" cell structures:
- What functions the insurgents need to be able to perform.
- What equipment is needed to perform those functions.
- How many people are required to employ the required equipment.
- The number of vehicles in relation to the people needed to drive them or the people and equipment that must be transported.
- Equipment associated with other equipment (for example, an aiming circle/goniometer used with a mortar or a day/night observation scope used with a sniper rifle).

6-24. Any relationship of independent local insurgent organizations to regional or national insurgent structures may be one of affiliation or dependent upon a single shared or similar goal. These relationships are generally fluctuating and may be fleeting, mission-dependant, or event- or agenda-oriented. Such relationships can arise and cease due to a variety of reasons or motivations.

6-25. When task-organizing insurgent organizations, guerrilla units might be subordinate to a larger insurgent organization. However, they might be only loosely affiliated with an insurgent organization of which they are not a part. A guerrilla unit or other insurgent organization might be affiliated with a regular military organization. A guerrilla unit might also become a subordinate part of an OPFOR task organization based on a regular military unit.

GUERRILLA ORGANIZATIONS

6-26. Guerrilla organizations may be as large as several brigades or as small as a platoon or independent hunter-killer (HK) teams. Even in the AFS organizational directories, some guerrilla units were already reconfigured as HK units. In the fighting force structure represented in an OPFOR OB, some additional guerrilla units may become task-organized in that manner.

6-27. The structure of a guerrilla organization depends on several factors. These might include the physical environment, sociological demographics and relationships, economics, and/or support available from external organizations and countries. A guerrilla organization might be affiliated with forces from countries other than the state with which it is in conflict or other organizations external to the state sovereignty in contest. Some guerrilla organizations may constitute a paramilitary arm of an insurgent movement, while others may pursue guerrilla warfare independently from or loosely affiliated with an insurgent organization. Figure 6-2 on page 6-6 shows the baseline organization for guerrilla HK company as an example.

Chapter 6

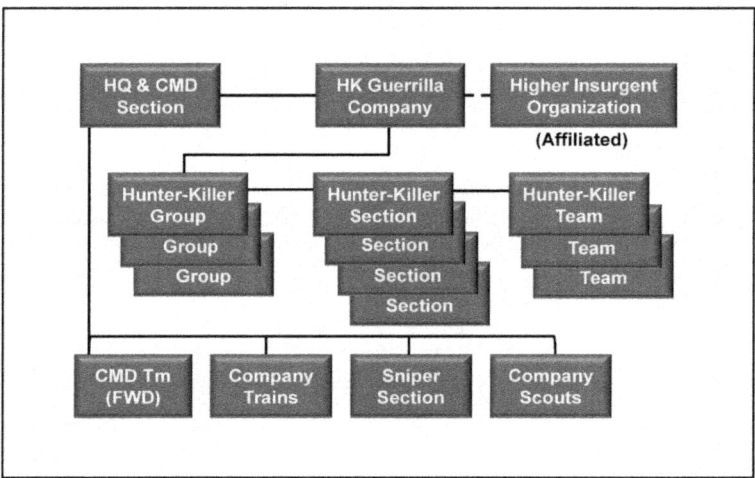

Figure 6-2. Guerrilla hunter-killer company (example)

6-28. Compared to insurgent organizations as a whole, guerrilla organizations have a more military-like structure. Within this structure, guerrilla organizations have some of the same types of weapons as a regular military force. The guerrilla organization contains weapons up to and including 120-mm mortars, antitank guided missiles (ATGMs), and man-portable air defense systems (MANPADS), and can conduct limited mine warfare and sapper attacks. Other examples of equipment and capability the guerrillas have in their organizations that the insurgents generally do not have are—

- 12.7-mm heavy machineguns.
- .50-cal antimateriel rifles.
- 73-, 82-, and 84-mm recoilless guns.
- 100-mm and 120-mm mortars.
- 107-mm multiple rocket launchers.
- 122-mm rocket launchers.
- Global positioning system (GPS) jammers.
- Signals reconnaissance capabilities.

CRIMINAL ORGANIZATIONS

6-29. Criminal organizations are normally independent of nation-state control, and large-scale organizations often extend beyond national boundaries to operate regionally or worldwide. Individual criminals or small-scale criminal organizations do not have the capability to adversely affect legitimate political, military, and judicial organizations. However, large-scale criminal organizations do. The weapons and equipment mix varies based on type and scale of criminal activity. Criminal organizations can appear similar to the characteristics of a paramilitary organization. Figure 6-3 shows an example of a large-scale criminal organization.

Hybrid Threat Organizations

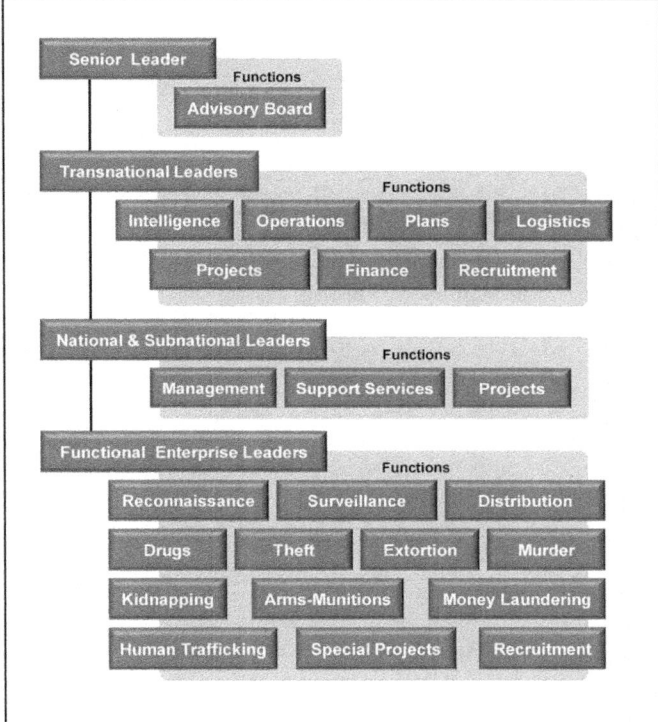

Figure 6-3. Large-scale criminal organization (example)

6-30. By mutual agreement, or when their interests coincide, criminal organizations may become affiliated with other actors, such as insurgent or guerrilla forces. Insurgents or guerrillas controlling or operating in the same area can provide security and protection to the criminal organization's activities in exchange for financial assistance, arms and materiel, or general logistical support. On behalf of the criminal organization, guerrilla or insurgent organizations can conduct—

- Diversionary actions.
- Reconnaissance and early warning.
- Money laundering, smuggling, or transportation.
- Civic actions.

Their mutual interests can include preventing U.S. or local government forces from interfering in their respective activities.

6-31. At times, criminal organizations might also be affiliated with nation-state military and/or paramilitary actors. In time of war, for instance, the state can encourage and materially support criminal organizations to commit actions that contribute to the breakdown of civil control in a neighboring country.

6-32. Criminal organizations may employ criminal actions, terror tactics, and militarily unconventional methods to achieve their goals. They may have the best technology, equipment, and weapons available, simply because they have the money to buy them. Criminal organizations may not change their structure in wartime, unless wartime conditions favor or dictate different types of criminal action or support activities.

6-33. Criminal organizations may conduct civic actions to gain and maintain support of the populace. A grateful public can provide valuable security and support functions. The local citizenry may willingly provide ample intelligence collection, counterintelligence, and security support. Intelligence and security can also be the result of bribery, extortion, or coercion.

HYBRID RELATIONSHIPS

6-34. The HT is a composite of many different groups. These groups will often have no standard, readily identifiable organizational relationship. What brings together the capabilities and intent of the components of the HT is a common purpose, typically opposition to U.S. goals. This unity of purpose can even bring together groups that normally would be fighting among themselves.

6-35. *Affiliated* organizations are cooperating toward a common goal despite having no formal command or organizational relationship. Affiliated organizations are typically nonmilitary or paramilitary groups such as criminal cartels, insurgencies, terrorist cells, or mercenaries.

6-36. Those irregular forces operating in a military unit's AOR that the unit may be able to sufficiently influence to act in concert with it for a limited time are affiliated forces. No "command relationship" exists between an affiliated organization and the unit in whose AOR it operates. In some cases, affiliated forces may receive support from the military unit as part of the agreement under which they cooperate.

Appendix A
Scenario Blueprints

A training scenario blueprint is a pictorial and textual representation of the results of task and countertask analysis. A blueprint is presented as a course of action (COA) sketch with accompanying text. The actions and entities depicted establish the necessary full spectrum training conditions that provide the opportunity to accomplish training objectives.

SCENARIO BLUEPRINT CONCEPT

A-1. The purpose of a scenario blueprint is to serve as a framework of circumstances and situations that will provide the appropriate full spectrum training for leaders and units. Blueprints should not be considered rigid templates and must be modified to accommodate training desired tasks. They are the result of the exercise design process and provide a starting point for battlefield geometry, potential application of the operational variables (PMESII-PT), logical hybrid threat forces, and training conditions. A training scenario blueprint provides an example of the key circumstances, situations, events, and actions of a training event with COA sketches and text.

A-2. Blueprints are intended to be used in conjunction with TC 7-101, *Exercise Design*, to fully develop comprehensive full spectrum training events. A scenario blueprint provides a basis to assess the resources required to establish the conditions of the operational environment (OE) needed to adequately challenge the training tasks based on task and countertask analysis. Once the desired training unit tasks are determined, exercise planners must design an opposing force (OPFOR) to conduct the appropriate countertasks. A training venue may or may not be resourced with all the personnel, equipment, or facilities necessary to create the OE conditions required to fully train units on desired tasks. In the cases where the resources identified as necessary to adequately challenge training tasks are not available, the commander and training developers must assess training risk and develop mitigation strategies. Using TC 7-101 provides a tool for commanders and trainers to determine how much training risk is being accepted and what alternatives are available.

A-3. The geostrategic setting for the scenario—commonly known as the "road to war"—is created by exercise planners after all relevant conditions are selected. This strategic setting provides a logical framework for understanding the training OE. It should also reflect the unique character of the available training geography and local requirements. The actual or fictitious adversaries, the description of motivation for activities, and other elements that establish the logic of the OE conditions are added as needed once a training scenario that challenges the training tasks is completed. The identity and motivations of the adversary (real or fictitious) provide a context to the scenario that is necessary but secondary to establishing conditions to adequately challenge training tasks.

EXERCISE DESIGN

A-4. To start the exercise design process the training unit commander selects his training objectives and the operational theme for his training event. The operational theme may be major combat operations (MCO), irregular warfare (IW), or a combination. The training objectives are broken down into unit and leader tasks to be trained and the commander's assessment of the unit's current training state. An example

Appendix A

set of tasks for a heavy brigade combat team (HBCT) conducting offensive operations in a full-spectrum setting might be—
- Conduct an Attack.
- Conduct Security Operations.
- Conduct Lethal and Nonlethal Fire Support.
- Conduct Mobility Operations.
- Conduct Information Operations.
- Conduct Humanitarian and Civic Assistance.
- Protect Critical Assets.

A-5. Exercise planners examine the training tasks and conduct a countertask analysis to determine what is needed to challenge the tasks at the appropriate level. The countertask analysis includes all the elements of the OE that may be necessary to challenge the tasks. A set of countertasks that challenges the tasks in the above example might be—
- Conduct Maneuver Defense.
- Conduct Disruption.
- Conduct Actions on Contact.
- Conduct Counterreconnaissance.
- Prepare Obstacles.
- Conduct Information Warfare (INFOWAR).
- Conduct Insurgent Operations.

A-6. With the task and countertask analysis complete, the exercise planners choose an OE, an OPFOR, and a set of COAs that provides the countertasks as a challenging set of conditions. This methodology is explained in TC 7-101. That TC provides the tools to properly scale the OPFOR and develop the needed OE conditions to support the countertasks developed. TC 7-101 provides the step-by-step process to fill out the blueprint framework with the detailed planning and resources needed to conduct the countertasks.

A-7. Typically, the countertasks suggest a COA. There is not one right solution, so long as the solution chosen challenges the things the unit is trying to accomplish. The sources for OPFOR COAs are the TC 7-100 series. TC 7-101 contains the OPFOR Tactical Task List, which is the source for countertasks. Continuing the example, the exercise planners see that the selected countertasks support a particular OPFOR COA. Figure A-1 shows an example of the COA sketch and text (COA statement) that go with the OPFOR countertasks listed above.

SCENARIO BLUEPRINT EXAMPLES

A-8. This appendix provides six examples to explain how scenario blueprints are designed and used. The first example, which describes the process in detail, is a continuation of the exercise design example above. The remaining five examples are provided for illustrative purposes. The six examples each individually focus on an emphasized aspect of full spectrum operations. Three of the examples do so within an MCO operational theme and three within an IW operational theme. All six examples are based on an HBCT conducting tasks relevant to the operational theme. Each scenario blueprint provides for an emphasis on training one component of full spectrum operations with the other components present but not the training focus. However, training tasks associated with the non-emphasis components will be included in the design and conditions established that provide the opportunity to train those designated tasks.

MCO BLUEPRINT EXAMPLES

A-9. The following are three examples of scenario blueprints with an MCO operational theme. One example has an offensive emphasis, one a defensive emphasis, and the other a stability operations emphasis for the training unit.

MCO Blueprint Example 1: Offensive Emphasis Blueprint

A-10. For illustrative purposes, this blueprint is based on the example of training unit tasks and OPFOR countertasks in paragraphs A-4 and A-5 above. In this case, the OPFOR chooses to conduct a maneuver defense involving a combination of regular military and insurgent forces. (See figure A-1.)

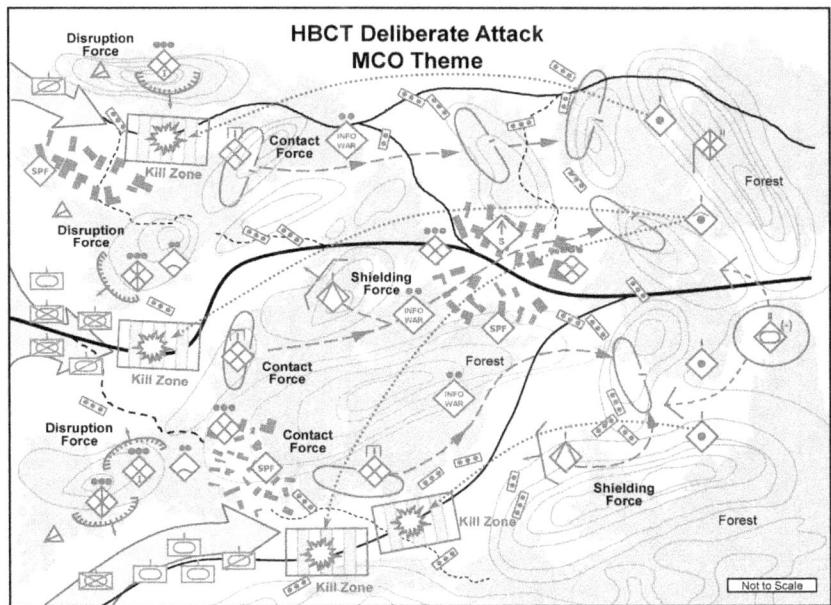

Figure A-1. MCO Blueprint COA sketch (example 1)

A-11. **Course of Action: OPFOR Maneuver Defense.** The OPFOR disrupts command and control (C2) and security forces to deny effective situational understanding and provide freedom of maneuver to the contact force. The disruption force consisting of an infantry-based detachment including special-purpose forces (SPF) teams and affiliated insurgent forces—

- Fixes the reconnaissance squadron.
- Conducts deception, electronic warfare, and perception management.
- Forces the early deployment of the combined arms battalions.
- Destroys intelligence, surveillance, and reconnaissance (ISR) assets.

A-12. The contact force consists of an infantry-based detachment with supporting artillery and information warfare (INFOWAR) assets. It delays the combined arms detachments and forces the HBCT to slow and deploy its fires, sustainment and C2 assets in areas vulnerable to attack by OPFOR fires and the disruption force. The shielding force consists of an antitank-based detachment with supporting artillery and INFOWAR assets. It fixes the two combined arms battalions by conducting attack by fire and permits the contact force to conduct a retrograde to pre-planned battle positions. When this retrograde is complete, the shielding force becomes the new contact force, and the maneuver is repeated until the HBCT's attack is culminated and is vulnerable to counterattack.

A-13. By this point, an initial depiction of the exercise scenario and the sequence of events has emerged. However, the scenario has to be constructed to accommodate training area terrain and constraints. The

Appendix A

scenario developer fits the depiction to the terrain and timeline for the training event. The end result is a training event that contains all the components of full spectrum operations in a realistic OE.

MCO Blueprint Example 2: Defensive Emphasis Blueprint

A-14. In this blueprint, the OPFOR chose to use a dispersed attack to overcome some of the U.S. advantages in ISR, close air support, and other stand-off fires. (See figure A-2.) If the training tasks called for the training unit to operate without those enablers, the OPFOR might have chosen to conduct an integrated attack.

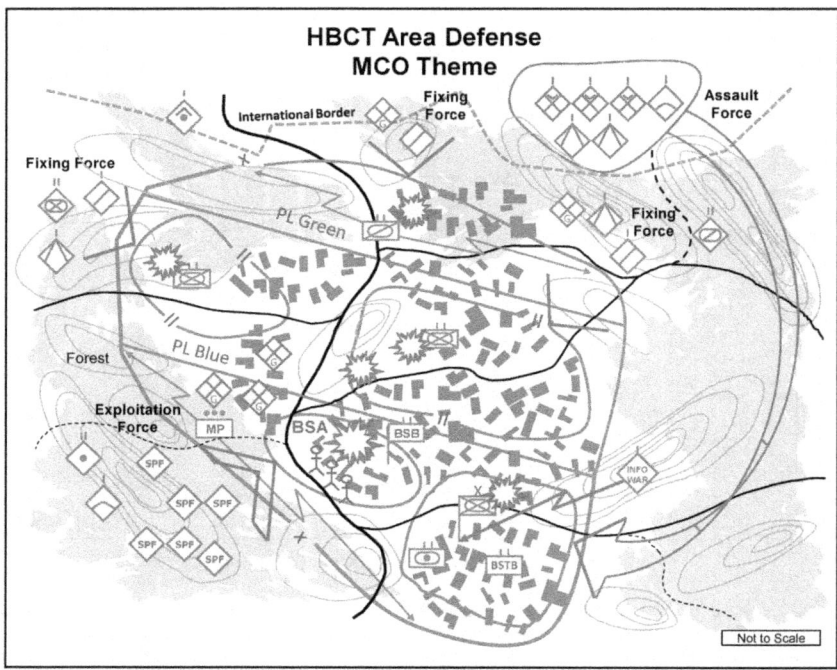

Figure A-2. MCO Blueprint COA sketch (example 2)

A-15. **Course of Action: OPFOR Dispersed Attack.** The OPFOR disrupts C2 and security forces to prevent effective response to other actions and deny effective situational understanding. The disruption force consisting of a reconnaissance battalion, two antitank batteries, an INFOWAR unit, guerrillas, and a supporting rocket launcher battery—
- Fixes the reconnaissance squadron.
- Conducts deception, electronic warfare, and perception management.
- Blocks quick reaction force (QRF) and reserve routes.
- Destroys ISR assets.

A-16. The OPFOR fixes the two combined arms battalions by conducting attacks with three combined arms detachments. The mission of the fixing force is to prevent either combined arms battalion from repositioning significant combat power to protect the HBCT's C2 and sustainment assets from attack. The OPFOR conducts an air assault to destroy HBCT C2 and fires assets in order to permit effective action by the exploitation force. The air assault consists of a battalion-size infantry and antitank detachment

Scenario Blueprints

augmented by INFOWAR elements. The OPFOR destroys the HBCT's sustainment capability by attacking the brigade support area (BSA) with an exploitation force consisting of six SPF teams, an artillery battalion, an air defense battery, and affiliated guerrilla forces.

MCO Blueprint Example 3: Stability Operations Emphasis Blueprint

A-17. In this blueprint, none of the countertasks chosen include conduct of combat operations by conventional, regular military forces. Therefore, the OPFOR COA focuses on guerrilla operations (see figure A-3).

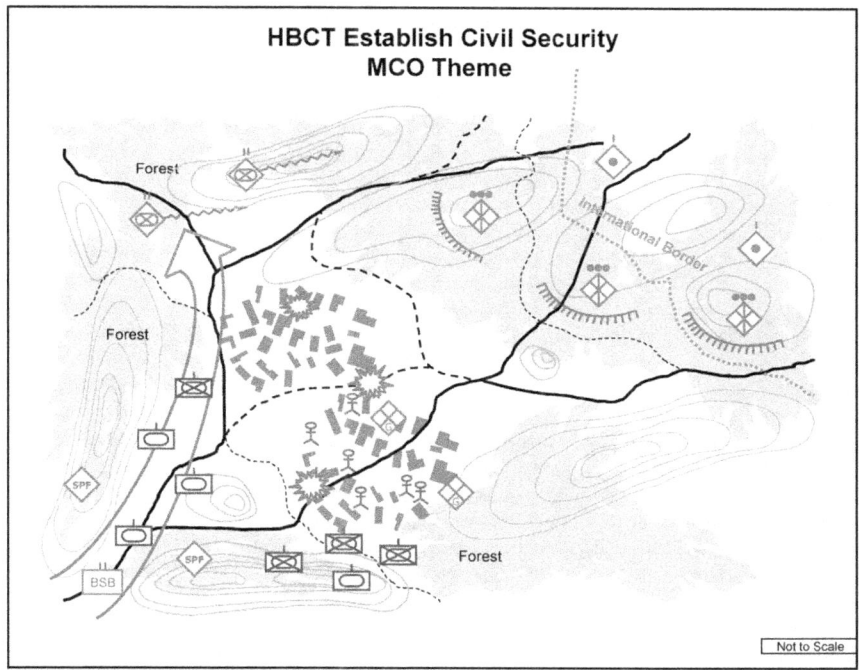

Figure A-3. MCO Blueprint COA sketch (example 3)

A-18. **Course of Action: OPFOR Guerrilla Operations.** The OPFOR disrupts C2 and security forces to deny effective situational understanding and provide freedom of maneuver to the guerrilla force. The disruption force, including guerrilla elements and SPF teams—

- Fixes the reconnaissance squadron (not shown).
- Conducts deception, electronic warfare, and perception management.
- Prevents effective sustainment.
- Prevents the creation of stable civil functions.

A-19. OPFOR conventional force remnants occupy defensive positions or move toward sanctuary— international border or rugged terrain. The conventional force near the international border is prepared to provide aid to guerrilla elements as needed.

Appendix A

IW BLUEPRINT EXAMPLES

A-20. The following are three examples of scenario blueprints with an IW operational theme. (See figures A-4 through A-6.) One example has an offensive emphasis, one a defensive emphasis, and one a stability operations emphasis for the training unit.

IW Blueprint Example 1: Offensive Emphasis Blueprint

A-21. In this blueprint, the countertasks required are executed primarily by regular military forces conducting an area defense, with affiliated guerrilla elements acting as part of the disruption force. (See figure A-4.)

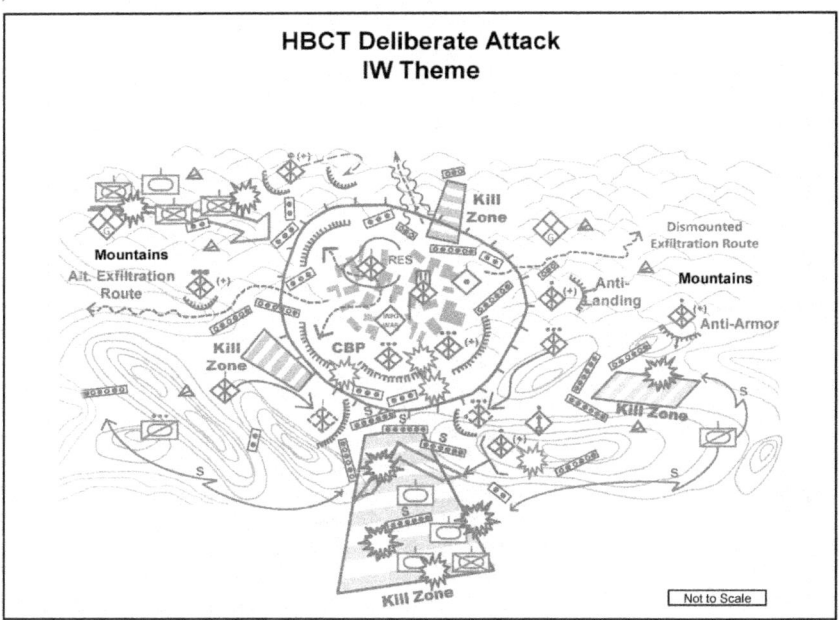

Figure A-4. IW Blueprint COA sketch (example 1)

A-22. **Course of Action: OPFOR Area Defense.** The OPFOR disrupts C2 and security forces to deny effective situational understanding and provide freedom of maneuver to the main defense force. The disruption force outside the complex battle position (CBP) of the main defense force—

- Fixes the reconnaissance squadron (only part of which is shown in figure A-4).
- Conducts deception, electronic warfare, and perception management.
- Forces the early deployment of the combined arms battalions; and destroys ISR assets.

A-23. The main defense force consists of an infantry-based detachment with supporting engineer and air defense assets. It defends from a CBP and protects OPFOR C2, INFOWAR, fires, and sustainment from enemy attack

A-24. The OPFOR reserve blocks the enemy shaping force and permits freedom of maneuver to the counterattack and main defense forces. The counterattack force blocks the decisive force and permits the protected force (C2, INFOWAR, fires, and sustainment assets) to destroy the HBCT's sustainment and C2 assets and then exfiltrate to remain viable for future battles.

IW Blueprint Example 2: Defensive Emphasis Blueprint

A-25. In this blueprint, the countertasks are executed by guerilla forces attacking a perimeter defense. (See figure A-5.)

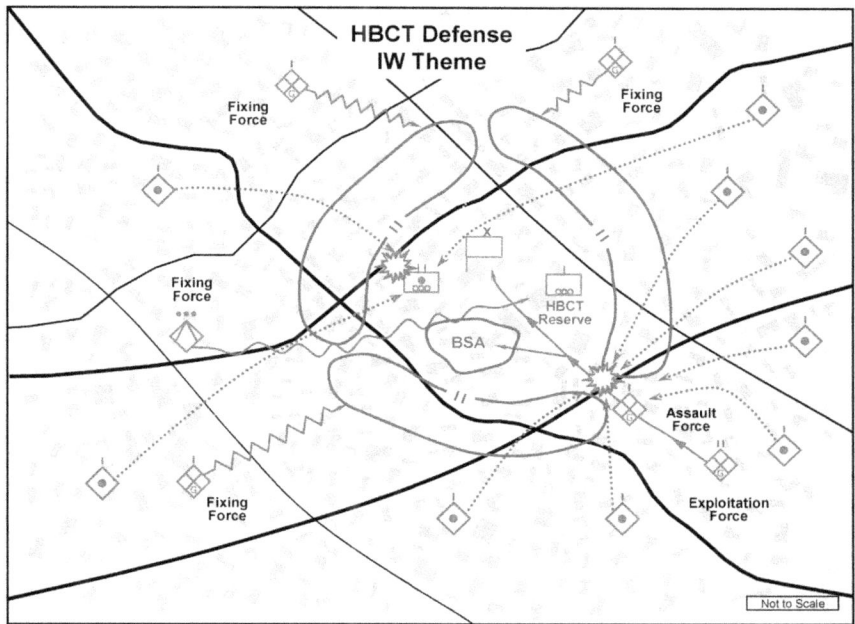

Figure A-5. IW Blueprint COA sketch (example 2)

A-26. **Course of Action: OPFOR Integrated Attack.** The OPFOR disrupts C2 and security forces to prevent effective response to other actions and deny effective situational understanding. The disruption force (not shown)—

- Fixes the reconnaissance squadron.
- Conducts deception, electronic warfare, and perception management.
- Blocks reserve routes.
- Destroys ISR assets.

A-27. The OPFOR fixes the three combined arms battalions by conducting attacks with multiple guerrilla elements with supporting fires and antitank assets. The mission of the fixing force is to prevent any of the combined arms battalions from repositioning significant combat power to protect the HBCT's C2 and sustainment assets from attack. The OPFOR conducts an assault with a guerrilla element to destroy HBCT C2 and fires assets, to permit effective action by the exploitation force. The OPFOR destroys the HBCT's sustainment capability by attacking the BSA with an exploitation force consisting of guerrilla forces.

Appendix A

IW Blueprint Example 3: Stability Operations Emphasis Blueprint

A-28. In this blueprint, the countertasks required are those that oppose stability action. Some of these countertasks will be executed by noncombatant actors opposed to creation of a stable environment due to political, ethnic, or religious motivations. (See figure A-6.)

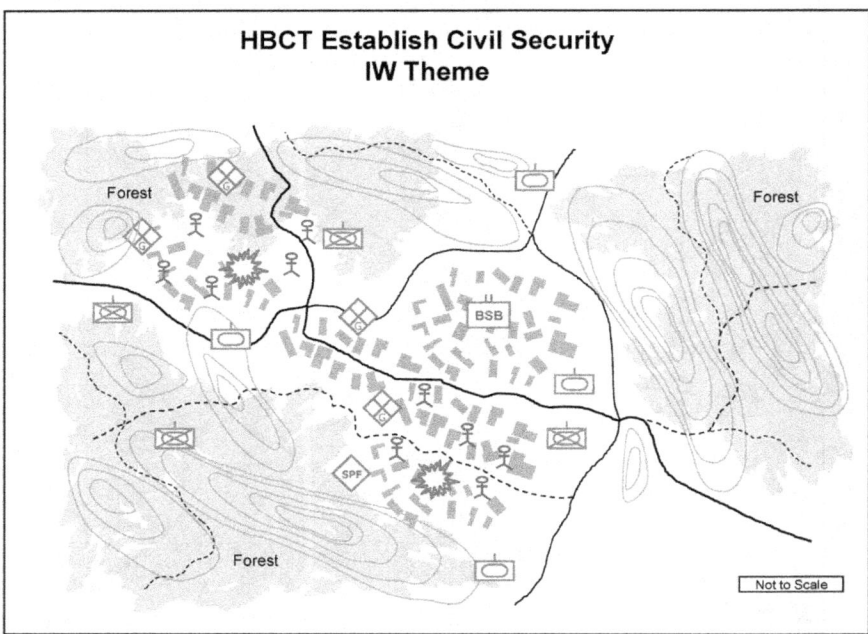

Figure A-6. IW Blueprint COA sketch (example 3)

A-29. **Course of Action: OPFOR Guerrilla Operations.** The OPFOR disrupts C2 and security forces to deny effective situational understanding and provide freedom of maneuver to the guerrilla force. The disruption force fixes the reconnaissance squadron (not shown); conducts deception, electronic warfare, and perception management; prevents effective sustainment; and prevents the creation of stable civil functions.

Glossary

SECTION I – ACRONYMS AND ABBREVIATIONS

AFS	administrative force structure
AGL	automatic grenade launcher
AKO	Army Knowledge Online
alt	alternate
AOR	area of responsibility
AP	antipersonnel
APOD	aerial part of debarkation
APOE	aerial part of embarkation
AT	antitank
ATGM	antitank guided missile
BSA	brigade support area
BSB	brigade support battalion
BSTB	brigade special troops battalion
C2	command and control
C3D	camouflage, concealment, cover, and deception
cal	caliber
CBP	complex battle position
CI	counterintelligence
CMD	command
COA	course of action
CSOP	combat security outpost
CTID	Contemporary Operational Environment and Threat Integration Directorate
DOD	Department of Defense
DODD	Department of Defense Directive
EW	electronic warfare
FM	field manual
FMFRP	Fleet Marine Force Reference Publication
FWD	forward
G	guerrilla (in unit symbol)
GPS	global positioning system
HBCT	heavy brigade combat team
HK	hunter-killer
HQ	headquarters
HRO	humanitarian relief organization
HT	Hybrid Threat (for training)

Glossary

HUMINT	human intelligence
I	insurgent (in unit symbol)
IA	information attack
IED	improvised explosive device
INFOWAR	information warfare
ISR	intelligence, surveillance, and reconnaissance
IW	irregular warfare
JP	joint publication
LOC	line of communications
MANPADS	man-portable air defense system
MCO	major combat operations
METL	mission essential task list
mm	millimeter
OB	order of battle
OE	operational environment
OP	observation post
OPFOR	opposing force
PMESI-PT	political, military, economic, social, information, infrastructure, physical environment, and time (*see also* PMESII-PT under terms)
QRF	quick reaction force
RES	reserve
RISTA	reconnaissance, intelligence, surveillance, and target acquisition
ROE	rules of engagement
SATCOM	satellite communications
SBP	simple battle position
SHC	Supreme High Command
SPF	special-purpose forces
SPOD	sea port of debarkation
SPOE	sea port of embarkation
TC	training circular
TRADOC	U.S. Army Training and Doctrine Command
TRISA	TRADOC G-2 Intelligence Support Activity
U.S.	United States
USMC	United States Marine Corps
W	weapons platoon (in unit symbol)
WMD	weapons of mass destruction

SECTION II – TERMS

hybrid threat
> The diverse and dynamic combination of regular forces, irregular forces, and/or criminal elements all unified to achieve mutually benefitting effects.

operational environment
> A composite of the conditions, circumstances, and influences that affect the employment of capabilities and bear on the decisions of the commander. (JP 3-0)

operational variables
> Those interrelated aspects of an operational environment, both military and nonmilitary, that differ from one operational environment to another and define the nature of a particular operational environment. The eight operational variables are political, military, economic, social, information, infrastructure, physical environment, and time. *See* PMESII-PT.

PMESII-PT
> A memory aid for the operational variables used to describe an operational environment: political, military, economic, social, information, infrastructure, physical environment, and time.

This page intentionally left blank.

References

Department of the Army Forms

The DA Form is available on the Army Publishing Directorate web site (www.apd.army.mil).

DA Form 2028, *Recommended Changes to Publications and Blank Forms*.

Documents Needed

These documents must be available to the intended users of this publication.

FM 1-02, *Operational Terms and Graphics*. 21 September 2004.

JP 1-02, *Department of Defense Dictionary of Military and Associated Terms*. Available online: http://www.dtic.mil/doctrine/jel/doddict/

TC 7-101, *Exercise Design*. 26 November 2010.

Readings Recommended

These sources contain relevant supplemental information.

FM 7-100.1, *Opposing Force Operations*. 27 December 2004.

FM 7-100.4, *Opposing Force Organization Guide*. 3 May 2007. Associated online organizational directories, volumes I-IV, available on TRADOC G2-TRISA Website at https://www.us.army.mil/suite/files/19296289 (AKO access required). Associated *Worldwide Equipment Guide*, volumes 1-3, available on TRADOC G2-TRISA Website at https://www.us.army.mil/suite/files/14751393 (AKO access required).

Sources Used

These are sources quoted or paraphrased in this publication.

DODD 2310.01E, *The Department of Defense Detainee Program*. 5 September 2006.

DODD 3000.7, *Irregular Warfare (IW)*. 1 December 2008.

Irregular Warfare: Countering Irregular Threats Joint Operating Concept, Version 2.0. 17 May 2010.

USMC FMFRP 12-18, *Mao Tse-tung on Guerrilla Warfare*. 5 April 1989.

This page intentionally left blank.

Index

Entries are by paragraph number unless page (p.) or pages (pp.) is specified. After a page reference, the subsequent use of paragraph reference is indicated by the paragraph symbol (¶).

A

access limitation, 4-24–4-26
action force or element, 5-19
action functions, 5-19–5-31
adapting by function, 5-15–5-16
adaptive operations, p. 3-2, ¶3-7, 3-8, 4-1, 4-16–4-22
allow no sanctuary, 4-47–4-50
area defense, 5-36–5-37
　example, A-19–A-20
assault force or element,
　as action function, 5-20, 5-22
　as enabling function, 5-34

B

battle position,
　complex, 5-8–5-11, 5-37, 5-41–5-46
　simple, 5-9
breach element, 5-22, 5-33

C

casualties, 4-32–4-34
cause politically unacceptable casualties, 4-32–4-34
change the nature of conflict, 4-42–4-46
combatants. See enemy combatants.
complex battle position, 5-8–5-11, 5-37, 5-41–5-46
computer warfare, 3-27
contact force, 5-20
control tempo, 4-27–4-31
counterattack force, 5-58
criminal organizations, 2-31–2-35, 6-29–6-33
　affiliation with guerrillas or insurgents, 2-33, 6-30
　affiliation with nation-states, 2-35, 6-31
　terror tactics, 3-16, 6-32
cultural standoff, 5-11

D

deception,
　as element of INFOWAR, 3-27, 5-6
　as enabling function, 5-56
deception force or element, 5-56
disbursed attack,
　examples, A-14, A-22
disruption, 5-39
disruption force, p. 5-7, ¶5-39

E

electronic warfare, 3-27, 5-6
enabling force or element, 5-32
enabling functions, 5-32–5-58
enemy combatants, 2-22–2-34
　lawful, 2-22, p. 2-4
　unlawful, 2-22, p. 2-4
exercise design, A-1–A-7
exploitation force, 5-20
extraregional power, 3-3
　principles of operation versus one, 4-23–4-54

F

fixing, 5-47–5-55
fixing force or element, 5-33, p. 5-7
functional tactics, 5-17–5-58
　action functions, 5-19–5-31
　enabling functions, 5-32–5-58

G

guerrilla. See also guerrilla warfare.
　definition, p. 2-4, ¶2-27
guerrilla operations
　examples, A-16, A-24
　hunter-killer teams, company, 6-26
　in insurgent organization, 6-25–6-26
　organizations, 6-26–6-28
　paramilitary, 2-24
guerrilla warfare, 2-27
　by irregular forces, 2-19
　Maoist example, pp. 2-5–2-6

H

hunter-killer,
　company, 6-26–6-27
　teams, 6-26
hybrid threat(s), pp. 1-1–2-7.
　See also Hybrid Threat (for training).
　adaptation, 1-11–1-14
　components, pp. 2-1–2-7
　concepts, p. 1-1–1-3
　definition, pp. iv, 1-1
　emergence of, pp. iv–v
　historical examples, 1-3
　major combat operations, p. v
　multiple threats, p. v
　transitions, 1-15–1-18
　WMD capability, 1-1
Hybrid Threat (for training), p. 3-1, pp. 3-1–6-8
　administrative force structure, 6-5
　affiliated relationships, 6-35–6-36
　criminal organizations, 6-29–6-33
　functional tactics, 5-17–5-58
　guerrilla organizations, 6-27–6-28
　insurgent organizations, 6-19–6-26
　instruments of power, p. 3-1, ¶3-1, 3-5, 3-11–3-13. p. 3-4
　internal security forces, 6-4, 6-6, 6-10–6-15
　operational designs, 3-3–3-10, 4-1–4-2
　operations, pp. 4-1–4-8
　order of battle, 6-1, 6-5
　organizations, pp. 6-1–6-8
　strategic goals, 3-3–3-10
　strategic operations, 3-1–3-19
　strategy, pp. 3-1–3-7

Index

Entries are by paragraph number unless page (p.) or pages (pp.) is specified. After a page reference, the subsequent use of paragraph reference is indicated by the paragraph symbol (¶).

Supreme High Command, 6-4, 6-7
tactics, pp. 4-1–4-8
task-organizing, 6-1–6-2, 6-25
weapons of mass destruction, 3-29–3-32, 5-21

I

information attack, 3-27
information warfare,
  computer warfare, 3-27
  deception, 3-27
  electronic warfare, 3-27
  elements of, 3-27
  information attack, 3-27
  in strategic operations, 3-12, 3-13
  operational shielding, 4-53
  physical destruction, 3-27
  perception management, 3-27
  protection and security measures, 3-27
  strategic INFOWAR, 3-20–3-28
  tactical INFOWAR, 5-5–5-6
INFOWAR. *See* information warfare.
insurgent. *See also* insurgency.
  definition, p. 2-4
  organizations, 2-25–2-26, 6-19–6-25
  paramilitary, 2-24
  terror tactics, 3-16
insurgency, 2-25
internal security forces, 6-4, 6-6, 6-10–6-15
irregular forces, 2-15–2-21, 5-3–5-4. *See also* criminal, guerrilla, or insurgent.
  compared to regular military forces, 2-17–2-21
irregular warfare, 2-15. *See also* irregular forces.
  scenario blueprint examples, A-17–A-24
  organizations, 6-3–6-5
IW. *See* irregular warfare.

L

lawful enemy combatant, 2-22, p. 2-4

M

main defense, 5-20

main defense force or element, 5-20
maneuver defense, example, A-11–A-12
military forces. *See* regular military forces.
mercenaries, 2-24. 2-29–2-30, 6-35
militia, 2-23, p. 2-5, ¶6-4, 6-16, 6-18

N

nation-state actors, 2-5–2-9
  core states, 2-6
  failed or failing states, 2-9
  rogue states, 2-8
  transition states, 2-7
nature of conflict, changing, 4-42–4-46
neutralize technological overmatch, 4-35–4-41
non-state actors, 2-10–2-12
  rogue actors, 2-11
  third-party actors, 2-12

O

operational designs, 3-3–3-10, 4-1–4-2
operational exclusion, 4-51–4-52
operational shielding, 4-53–4-54
operational variables, 2-3, p. A-1
operations. *See* Hybrid Threat, operations.
OPFOR. *See* opposing force.
opposing force, p. 3-1. *See also* Hybrid Threat (for training).
  administrative force structure, 6-5
  insurgent organizations, 6-19–6-26
  internal security forces, 6-4, 6-6, 6-10–6-15
  order of battle, 6-5, 6-22, 6-23, 6-26
  organization guide, 6-1–6-2, 6-4, 6-22
  organizations, pp. 6-1–6-8
  regular military forces, 6-3–6-5
  Supreme High Command, 6-4, 6-7

P

paramilitary forces
  definition, p. 2-4, ¶2-24
  non-state, 2-24
  nation-state, 2-24
perception management, 3-27, 5-6
physical destruction, as element of INFOWAR, 3-27
PMESII-PT. *See* operational variables.
police, 6-13–6-14
protected force, p. 5-4
protection and security measures, as part of INFOWAR, 3-27

R

raid, 5-22–5-31
regional operations, p. 3-2, ¶3-6–3-10, 4-1, 4-3–4-7
regular military forces, 2-13–2-14, 5-3–5-4, 6-3–6-5
  Air Defense Forces, 6-4
  Air Force, 6-4
  Army, 6-4
  compared to irregular forces, 2-17–2-21
  Internal Security Forces, 6-4, 6-6, 6-10–6-15
  Navy, 6-4
  organizations, 6-3–6-18
  Special-Purpose Forces, 6-4, 6-6–6-9
  Strategic Forces, 6-4
reserve, 5-58
reserves, 6-16–6-18

S

sanctuary, allowing none, 4-47–4-50
scenario blueprints, pp. A-1–A-8
  MCO examples, A-9–A-16
  IW examples, A-17–A-24
security, 5-40–5-46
shielding force, p. 5-20
special-purpose forces,
  in strategic operations, 3-12, 3-16
  SPF Command, 6-4, 6-1–6-9
  terror tactics, 3-16, 4-49

Index

Entries are by paragraph number unless page (p.) or pages (pp.) is specified. After a page reference, the subsequent use of paragraph reference is indicated by the paragraph symbol (¶).

SPF. *See* special-purpose forces.
States. *See* nation-state actors.
strategic goals, 3-1–3-10
strategic information warfare. *See* information warfare.
strategic operations, 3-1–3-19
   means, 3-11–3-13
   operational designs, 3-3–3-10
   strategic goals, 3-3–3-10
   targets, 3-14–3-16
   timeframe, 3-17–3-19
strategic preclusion, 3-29–3-30
strategy. *See* Hybrid Threat, strategy.
support element, 5-57
Supreme High Command, 6-4, 6-7
systems warfare, 5-12–5-14

T

tactics. *See* Hybrid Threat, tactics.
technological overmatch, neutralizing, 4-35–4-41
tempo, control of, 4-27–4-31
terror tactics, p. v. *See also* terrorist.
   by criminal organizations, 3-16, 6-32
   by Hybrid Threat, 4-45, 4-49, 4-50
   by insurgents, 3-16
   by irregular forces, 2-19
   by rogue actors, 2-22
   by rogue states, 2-8
   by SPF, 3-16, 4-49
   in strategic operations, 3-12, 3-16
   with WMD, 3-30
terrorist. *See also* terror tactics.
   definition, p. 2-4, ¶2-28
   group, 2-28, 6-35
   in strategic operations, 3-12, 3-14, 3-16
   paramilitary, 2-24
traditional warfare, 2-16
transition operations, p. 3-2, ¶3-7–3-10, 4-1, 4-8–4-15

U

unconventional warfare, 2-16

unlawful enemy combatant, 2-22, p. 2-4

W

warfare
   irregular warfare, 2-15
   traditional warfare, 2-16
   unconventional warfare, 2-16
weapons of mass destruction
   as action function, 5-21
   as enabling function, 5-21
   hybrid threat use of, 1-1, 2-36–2-39
   in strategic operations, 3-12

This page intentionally left blank.

TC 7-100
26 November 2010

By order of the Secretary of the Army:

GEORGE W. CASEY, JR.
General, United States Army
Chief of Staff

Official:

JOYCE E. MORROW
Administrative Assistant to
Secretary of the Army
1030801

DISTRIBUTION:
Active Army, Army National Guard, and United States Army Reserve: Not to be distributed; electronic media only.

PIN: 100556-000

www.ingramcontent.com/pod-product-compliance
Lightning Source LLC
Chambersburg PA
CBHW050238230526
45470CB00005B/2011